A SQUARE RIG HANDBOOK

Operations ★ Safety ★ Training ★ Equipment

Compiled by

Lieutenant-Commander F. J. M. Scott, MA, MNI, RN

With additional chapters by

Colin Mudie, C.Eng, FRINA,
Commander David Gay, MBE, MNI, RN (Retd)
and Andrew Cassell

First published 1992 by The Nautical Institute
202 Lambeth Road, London SE1 7LQ, UK.
Telephone: 071-928 1351

Typeset and printed in England by Silverdale Press, Silverdale Road, Hayes, Middlesex UB3 3BH.

ISBN 1 870077 11 3

Front cover: Lord Nelson *coming up the Needles Channel in a fresh breeze (MAX).*

A SQUARE RIG HANDBOOK
CONTENTS

	Page
Foreword	**10**
Preface	**11**
Acknowledgements	**12**

Chapter 1   **13**
SAILING SHIP DESIGN AND CONSTRUCTION
Design factors changed since the great days of sail
Factors affecting basic design
Construction

Chapter 2   **21**
MODERN EQUIPMENT

Annex 2A   **21**
RADAR & ELECTRONIC INSTRUMENTS
Radar arcs
Radar aerial height
Radar—daylight displays
Radar echoing area
Instruments—digital vs analogue

Annex 2B   **23**
SAILS
Sail cloths
Chafe protection
Sail care
Sails and UV
Sail repairs

Annex 2C   **27**
RUNNING AND STANDING RIGGING
Synthetic rope
Braidline
Coding lines
Stoppers in synthetic rope
Chain sheets
1×19 steel wire rope (SWR)
Standing rigging—stainless vs galvanised
Rod rigging
Terminals
Rigging screws
Shackles, eyebolts and eyeplates
Mousing shackles
Tufnol blocks
Special customised blocks
Sheaves
Marlingspike seamanship

Annex 2D   **31**
THE RIGGING SCHEDULE
Rigging schedule example

Annex 2E
MASTS AND SPARS   33
    Aluminium masts and spars
    Corrosion in aluminium spars
    Fatigue & stress
    Pole masts
    Mast exhausts
    Mast track vs parrals/hoops
    Wooden bowsprits

Annex 2F
SQUARE SAIL ROLLER FURLING GEAR   37
    Development background
    The new system
    The yard
    The furling line
    The roller or foil
    Attaching the sail
    Maintaining the roller
    Power operation
    Fixed yards

Chapter 3
MANOEUVRING UNDER SAIL   41
    Tacking
    Tacking—use of headsails
    Tacking—handling courses
    Wearing
    Wearing—intermast staysails
    Boxhauling
    Clubhauling
    Backing & filling
    Heaving-to
    Towing

Chapter 4
SAIL HANDLING AND TRIMMING   45
    Sail balance
    Balance & the spanker
    Trimming
    Shortening sail
    Sail size & sail handling
    Scandalising gaff sails
    Fanning the yards
    Square Sail 'slot effect'
    Bracing problems
    Reefing square sails

Annex 4A
RIGGING AND SAIL HANDLING VARIATIONS   49
    Jarvis clewlines
    Slablines
    Triangular courses
    Raffees
    'Flying' square sails
    Centre-furling square sails
    Paracee boom
    Bentinck boom
    Telescopic quarterbooms
    Winches various
    Running coils
    Bidevindstik

Annex 4B
FIXED AND HOISTING YARD SYSTEMS COMPARED 52
   Comparison table

Annex 4C
HANDLING ROLLER FURLING SQUARE SAILS 53
   Principles
   Problem areas
   Setting/furling a roller furling square sail

Chapter 5
AUXILIARY POWER AND THE SAILING VESSEL 56
   Propeller loading
   Propeller drag under sail—twin screw versus single screw
   Propeller drag under sail—shaft trailing
   Motoring to windward
   Motor-sailing
   Windage and manoeuvring in a confined area
   Single rudder—twin screw
   Single screw
   Variable-pitch propellers (VPP)
   Controllable-pitch propellers (CPP)
   Bow-thrusters
   Seaboat as a pusher
   Anchors

Chapter 6
HEAVY WEATHER 60
   Forecasting
   Heavy weather preparations
   Upper deck lifelines
   Running rigging—security
   Heavy weather shiphandling
   Use of oil
   Rounding-up in a storm
   Storm sails
   Scudding
   Heaving-to/lying-to
   Storm survival lessons
   Dismasting & jury rigs

Chapter 7
COASTAL PASSAGE PLANNING 64
   Windward ability
   Weather appreciation
   Strategy considerations-1
   Strategy considerations-2
   Embaying
   Anchorages & lee shores
   Conclusion

Chapter 8
THE RULE OF THE ROAD AND SHIPPING PROBLEMS UNDER SAIL   69
    Rule 3c 9b 10d(i) 10j 12 18a 18b 25 35c
    Rule 3f
    Rule 8f
    Rule 10c
    Rule 12
    Rule 12/13
    Rule 17a
    Rule 17c
    Rule 34b
    Shipping problems under sail
    Crossing—from starboard
    Crossing—from port
    Head-on situations
    Shipping problems—motor-sailing

Chapter 9
ROUTINES AND TRAINING   75
    Watchkeeping
    Joining routine
    Meal times
    Maritime pollution regulations
    Crew stations
    Training programme for cadets
    Training programme for afterguard
    Women at sea
    Foreign trainees

Chapter 10
SAFETY ONBOARD   79
    Safety philosophy
    Peer pressure
    Safety equipment
    Health & safety at work
    Climbing aloft in harbour
    Ratlines
    Backwires
    Futtock safety wires
    Hypothermia
    Watertight integrity and damage control
    Fire fighting
    Radio & radar transmission hazards (Radhaz)
    The galley
    The engineroom
    Anchors & cables
    Boat routines
    Hands to bathe
    First aid

Chapter 11
MANOVERBOARD UNDER SAIL AND UNDER POWER   85
    Manoverboard liferafts
    MOB markers
    Manoeuvres under sail
    Manoeuvres under power
    Controlling the seaboat
    Direct pick-up
    Medical treatment
    Outside assistance

Annex 11A
ONBOARD MOB LOCATION DEVELOPMENTS   90
  Trials
  Electronic location
  Whitbread race experience
  Future development

Annex 11B
NIGHT SEARCH AND LIFEBOAT CAPABILITIES   91
  Night search problems
  Lifeboat capability
  Electro-optics

Chapter 12
EMERGENCY HELICOPTER TRANSFERS   92
  Main types of SAR helicopter
  Helicopter search and rescue (SAR)
  Communications
  Time and distance
  Medevac preparations
  Pick-up methods
  Casualty considerations
  Recommended pick-up methods
  Hazards
  General points

Chapter 13
ABANDON SHIP AND LIFERAFT DRILLS   98
  Training
  Check list
  Liferaft drill
  Survival priorities
  Recovery by helicopter/lifeboat

Chapter 14
DYNAMIC STABILITY AND KNOCKDOWNS   101
  Incidents
  Knockdowns
  Downflooding
  Manoeuvres
  Countermeasures
  Actions after recovery from knockdown
  Conclusion

Annex 14A
SHIP STABILITY   107
  Standard stability abbreviations
  Useful formulae
  Stability curves
  Freeboard
  Caveats

Page

Chapter 15
LEGAL ASPECTS OF WORKING WITH YOUNG PEOPLE   **109**
     In loco parentis
     Responsibilities
     Problem areas
     Shore leave
     Crew exchanges

Appendix 1
EMERGENCY REFERENCE CARDS—EXAMPLES   **111**

Appendix 2
REFERENCES AND RECOMMENDED FURTHER READING   **115**

Appendix 3
USEFUL GERMAN & NORWEGIAN SAILING TERMS   **117**

Appendix 4
LIST OF ABBREVIATIONS   **118**

PHOTOGRAPHS

	Page
Lord Nelson coming up the needles channel in a fresh breeze (MAX)	Cover
Royalist — building (Author's collection)	16
Eagle — foremast mounted radar (Rosemary Mudie)	22
Alexander von Humboldt — main course yard (MAX)	24
Lord Nelson — mizzen topgallant staysail (MAX)	24
Dar Mlodziezy — course yard (MAX)	26
Lord Nelson — main topsail yard from aft (MAX)	28
Eagle — Chinese stopper (Morin Scott)	28
Royalist — spanker track (Author)	34
Royalist — fore topsail yard crane and hoisting arrangement (Author)	34
Lord Nelson — main topgallant yard from aft (MAX)	36
Lord Nelson — main topgallant yard from for'rard (MAX)	36
Centurion — stern view (Author)	48
Lord Nelson — foremast from aft (MAX)	48
Sorlandet — running coils for braces (Author)	51
Lord Nelson — fore topgallant and royal sails roller furled (MAX)	54
Belle Poule — fore topsail (Rosemary Mudie)	54
Royalist — stowing the mainsail (Author)	78
Sorlandet — stowing main topgallant — (Author)	80
Royalist — ratling bar detail — (Author)	84
Lord Nelson — manoverboard liferaft installation (MAX)	86
Hi-Line transfer (Royal Navy — RNAS Culdrose)	94

DIAGRAMS

	Page
Sail plan comparisons on the same hull (fig 1-1)	14
Roller furling gear layout — simplified view (fig 2F-1)	38
Tacking a brig (fig 3-1)	40
Wearing a brig (fig 3-2)	40
Boxhauling a brig (fig 3-3)	40
Channel passage ports (fig 7-1)	66
Rule of the road problems — Braced square — arc of manoeuvrability without bracing or tacking (fig 8-1)	70
Rule of the road problems — Braced up sharp for close hauled — arc of manoeuvrability without bracing or tacking (fig 8-2)	70
Rule of the road problems — Between sailing vessels (fig 8-3)	72
Rule of the road problems — Close hauled port tack (fig 8-4)	72
Rule of the road problems — Close hauled starboard tack (fig 8-5)	72
The Williamson turn (fig 11-11)	88
Captain vs *Monarch* GZ (fig 14-1)	102
Eurydice — open/closed GZ (fig 14-2)	102
Dynamic reserve — generic ship (fig 14-3)	103
GZ stability curves for contrasting generic vessels (fig 14-4)	104
Transverse stability diagram (fig 14-5)	104

FOREWORD

THE SAIL TRAINING COMMUNITY has rightly been described as a 'family', and it is thus both a great pleasure, and highly suitable, that a German should write the foreword to a book by a British author, who trained in a Norwegian square rigger, because this so typifies our multi-national spirit.

Written in a clear concise style, the reader will find that this is a comprehensive guide, which has benefited from feedback and contributions from a diverse group of individuals, coming from as far afield as Australia. The result is a well balanced and thoroughly up-to-date publication that avoids romantic nostalgia, and concentrates instead on the 'nuts and bolts' with a strong safety theme—and this is a viewpoint that I most strongly endorse.

I did serve in commercial deep-water sail, but as a former Captain of *Gorch Fock* I am very much aware that sail training cannot justify itself through mere sentiment. It does provide an excellent platform for basic seamanship training, but in both the naval and civilian worlds its main value in this modern machine age lies in its unique ability to foster the somewhat old-fashioned character virtues of courage, comradeship, and endurance—irrespective of race, creed, colour, or gender. We have been fortunate to see a real renaissance in sail training in the last 25 years, and continued success demands the high professional standards, modern outlook, and attention to detail, exemplified by the theme of this much-needed book. The written word can never be a substitute for experience and skill, however it can set out guidelines, and encourage breadth of mind, which is why I so wholeheartedly recommend this book to all professional seamen in our sail training family.

HORST-HELMUT WIND
Flottillen Admiral FGN

PREFACE

THIS BOOK has been compiled as part of the continuing initiative by The Nautical Institute to encourage the growth of square rig professionalism.

The work initially done by Captain R. M. Willoughby in his *Square Rig Seamanship* (published by The Nautical Institute in 1989) laid an excellent foundation, and the aim here is to expand on some of the issues that he raised, and to provide for those vessels that employ modern technology rather than purely traditional design and materials. The two books should be regarded as complementary, although inevitably there is some overlap.

All manuals have to be selective in their subject matter, because seamanship is such a vast subject, and this book is no exception. It is targeted not at trainees, but at those who serve (or wish to serve) as Master, Mate, or Boatswain, in a square rigger; consequently, it omits some of the basics, and as a baseline assumes that the reader has done at least one trip in such a vessel, and is familiar with the contents of the short trainees booklet that they all provide. It also does not cater for yachts, because they are already more than adequately covered by a myriad of yachting books, and the RYA examination system. However, it is expected that anyone sailing as an officer on larger vessels will also have yachting experience, and an RYA qualification—preferably Yachtmaster Offshore. As a result of the author's experience in producing aircraft tactical manuals, this book is broken down into short sections, so that the reader should be able to select the subject in which he/she is interested, more readily than with a normal index.

A reading list has been included to provide the serious reader, and those preparing for examination, with a guide to traditional—or specialist—manuals that expand on essential subject matter. The accident reports in that list should be considered as mandatory reading matter for everyone who goes to sea in sail in any position of authority.

Although written in English, time has been spent by the author, and/or other contributors, in vessels such as: *Sørlandet* and *Christian Radich* (Norway), *Gorch Fock* and *Alexander Von Humboldt* (Germany), *Amorina* (Sweden), *Eagle* (USA), *Esmeralda* (Chile), *George Stage* (Denmark), *Bounty* (Australia), and *Shabab Oman* (Oman), which should form a useful counterbalance to any parochial British bias.

F.J.M. SCOTT, MA, MNI
Lieutenant-Commander, RN

ACKNOWLEDGEMENTS

MY FATHER, Morin Scott, MBE, FNI, who originally suggested the need for this book, and Julian Parker, FNI (Secretary of The Nautical Institute), have both been a continual source of advice and encouragement.

No book as wide ranging as this can truly be the work of one person, and indeed much of the work has been collating and editing collective wisdom.

Colin Mudie, C.Eng, FRINA, who has been responsible for a number of sail training vessels (starting with *Royalist*), very kindly took the time and trouble to write the introductory chapter on ship design and construction, and also provided valuable technical advice and criticism. David Gay, MBE, a very experienced yachtsman, an RYA examiner, and a former Offshore Commander for the Sea Cadet Corps, produced the initial draft for the coastal passage planning chapter. Andrew Cassell, of the famous sailmakers Ratsey & Lapthorn Ltd, provided the very clear annex on sails. I hope that they forgive my editorial mauling.

Extensive information, or feedback, was also given by Hugh Munro of *Lord Nelson* (roller furling square sails), Barry Pickthall of *The Times* newspaper (manoverboard developments), Liz May of Hill Taylor Dickinson & Co, solicitors, (legal problems), and Stuart Welford of the RNLI (image intensifier trials).

In addition others have directly contributed by reading through various drafts, and/or providing much needed constructive criticism, suggestions, and advice. The reader will undoubtedly be particularly grateful to them for almost entirely purging the text of naval/aircrew terminology and slang:

Great Britain—Mark Kemmis-Betty; Josh Garner; John Hamilton; David Norman; Michael Willoughby; Len Mathews; Colin Hawksworth; Rex Barnes.
Australia—Ken Edwards.
New Zealand—Jenni Roberts.
Denmark—Barner Jespersen.
Germany—Hein Behn
USA—David Wood.

The cover photograph is courtesy of Max Mudie, and the remainder are as credited.

My thanks also go to Cathy Davis, who succeeded in turning my crude sketches into clear diagrams; and to Brian Mehl for the ColRegs diagrams and the cover artwork.

Finally, I should like to pay tribute both to the patience, and forbearance of the Captains, officers, and crews, of the sail training vessels in which I served as a cadet, and also to the challenge of the young trainees of today, who certainly do not believe in blind obedience, and who constantly remind me that in this game learning never stops.

Any errors or omissions are mine alone.

F.J.M.S.

Chapter 1

SAILING SHIP DESIGN & CONSTRUCTION

By Colin Mudie, C.Eng, FRINA

Introduction

TRADITIONAL ships were generally masterpieces of design and construction and given a similar set of design requirements the modern designer equipped with computers and modern engineering tools and weapons, such as finite element analysis, would have difficulty in matching them, never mind improving them.

However, the modern designer lives in a different world and has a number of quite differing circumstances to cater for when designing a new sailing vessel which may, with any luck, even have a superficial resemblance to those marvellous vessels of the past.

Design factors changed since the great days of sail

1. Auxiliary engines

The common use and indeed regulation requirement for sailing ships to have auxiliary engines has in itself practically eliminated one whole sailing function from the requirements of a new design, and introduced another. In previous sailing ships the need to be able to continue sailing when the winds fell light was a full and practical aspect of its operation. In many ships it led to over-sparring, and it certainly affected the hull form which had to be planned for minimum wetted surface. Serious light wind sailing is now largely accomplished by the propeller. The new sailing function is that of sailing to windward with the motor running. This effectively requires those elements of the rig which are used in this condition—such as the staysails—to be able to be sheeted much closer winded (see chapter 4). In addition the hull and rig may be operated into strong head sea conditions, which would have been unknown to the sail-only ship.

2. Other shipping

Generally speaking the sailing ships of the past operated amongst other vessels of similar speeds and manoeuvring characteristics, all influenced similarly by the ambient conditions. The modern sailing ship has to operate in very differing traffic, which may involve power driven vessels travelling at ten times her speed, which are likely to be substantially less manoeuvrable, and which may even carry right of way. The modern sailing vessel, therefore, must be planned with ample power and very good manoeuvrability, if she is to be operated safely amongst other shipping.

3. Harbours

Modern harbours are probably less crowded than they were in the heyday of sailing ships, but it has to be said that the berthing requirements of sailing ships are much less understood and appreciated than once they were. In consequence they are, often with the best motives, allocated berths which would have been considered as difficult and even inaccessible in the past. The only answer is for the designer to give them as much engine power as can be stomached by the sailormen on board, and to make them as manoeuvrable as possible in close-handling.

4. Voyage planning

Modern sail training ships have generally to operate to quite strict schedules, where departing crews have trains to catch, and arriving crews expect their ship to be exactly where—and when—it said in a brochure printed months before. Historically, this is a novel approach to sailing ship operation, but one which is of considerable importance to the success of current operation. In part it is covered by the use of the auxiliary

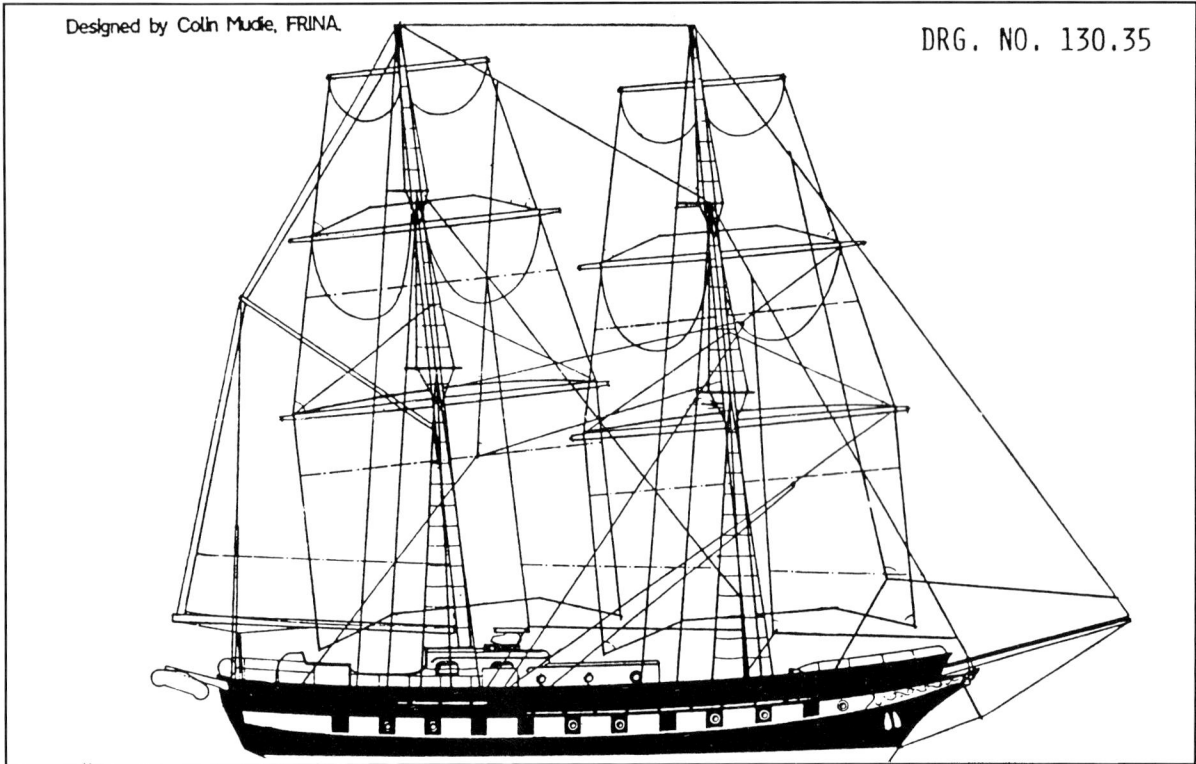

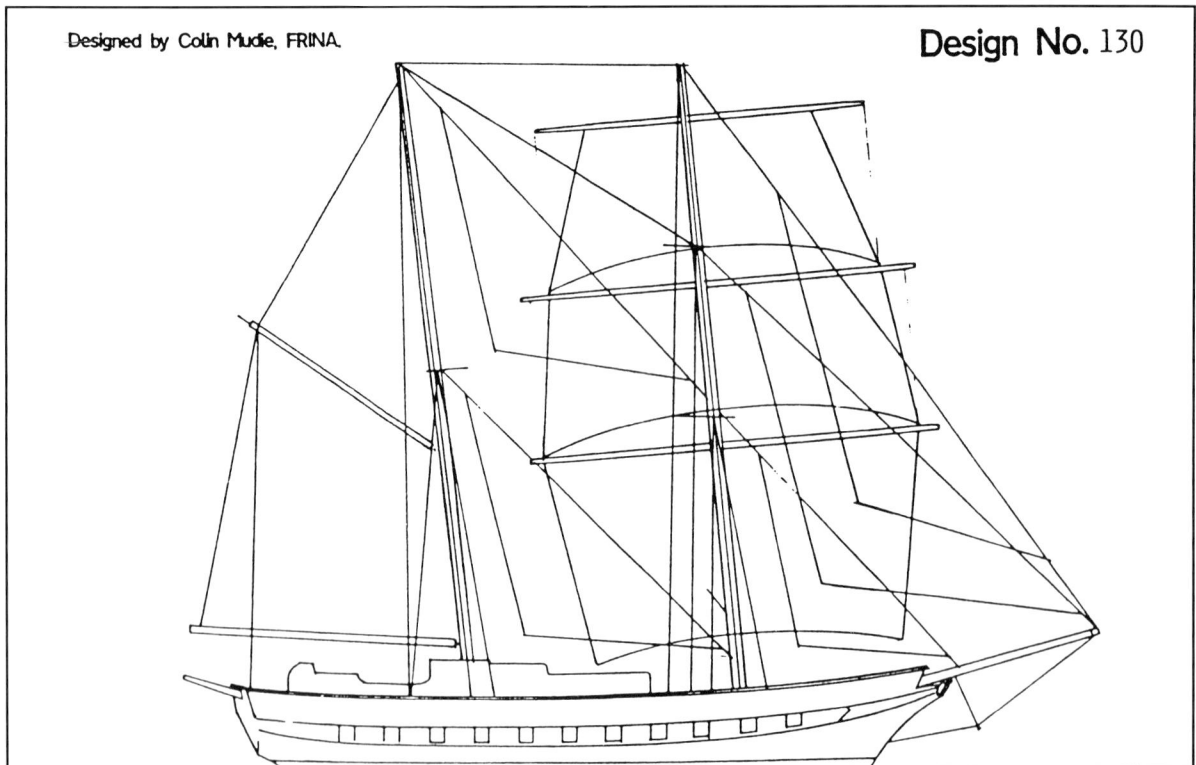

Fig 1-1 Sail plan comparisons on the same hull—the upper drawing shows Royalist *with her existing brig rig, and the lower one is the same hull with a brigantine rig. Note that much longer yards and much wider square sail plan is possible on the brigantine, while the yard length on the brig is constrained by the need to avoid fore and main yards fouling when braced in opposite directions (as in tacking).*

engines when required but such matters as refuelling, storing and down-time for maintenance have now become as important considerations for a sail training vessel as they are for a cruise liner.

5. Social changes

There are quite important social changes in the expectations of those on board which have to be taken into consideration. For instance, there are regulations regarding general living conditions, even affecting light and ventilation. The old measure for water tankage, a gallon a day per person, is now more like ten gallons per person. Another large social change lies in the accommodation provision for men, and women, who now put to sea together on a more or less equal number basis.

6. Materials

Changes of the materials available for construction have had less effect than the changes in the labour costs to work them. Aluminium masts with stainless fittings are now considerably cheaper than wooden spars with fittings which have to be made individually by blacksmiths. There is a slight gain in reduced weight aloft with aluminium, but it is largely costs which swing the balance away from wood. Modern sail cloths can be lighter as can the various items of rigging, but for sail training the differences probably lie more in first costs, and upkeep costs, rather than in any basic considerations which would affect the design (see annex 2B).

7. Fittings and equipment

Much as one might reflect on the days of simplicity at sea, it is plainly unseamanlike to operate a ship these days without the full range of position-finding and safety equipment. To a certain extent this approach extends into such items as electric cooking, food refrigeration, central heating and even air-conditioning, all of which can be considered as modern seamanlike aids to keeping the crew in top condition. Again, sewage now has to be treated, and garbage kept onboard (Marpol 73/78).

The modern sailing ship, therefore, has to carry an extensive set of auxiliary systems, and will normally have a generator running more or less continuously to support them.

8. Crew skills

It is possibly unlikely that modern professional crews will ever be able to accumulate the experience of those of the past, who—often from a very early age—spent comparatively little time ashore. The ships of the past had to be completely self-contained, and self-supporting for the full length of a voyage, which was accomplished with nothing but muscle, wind power and seamanship. However, it has to be said that in any case the modern regulating authorities have downgraded much of the earlier reliance on the professional skills for ship safety, and require each and every commercial sailing ship to be comparatively safe in its own right. It is said that they would not allow the *Cutty Sark* to sea again without substantial modifications to achieve these standards.

Factors affecting basic design

1. Stability

The transfer of responsibility from the owners and Master to the regulating authorities shows up principally in terms of stability, and downflooding. A large cargo-carrying sailing ship usually relied on added ballast, and the careful stowage of the cargo, for her stability on each and every voyage. This stability was often—in fact usually—less than would be approved today. This was not a failing of the basic understanding of all concerned, but a more sophisticated approach to seafaring. If the ship of the form required for cargo had been ballasted to modern requirements, her resultant stiffness would have given her a sharp roll, which would have been very hard on her lofty rig, and even harder on the seaman who had to go aloft to handle it. The responsibility for the

Royalist—building—as the skeleton takes shape it is a tremendous advantage to build fully under cover protected from the elements, as was done in this case over the winter of 1970/71 (Author's collection).

safety of the ship then had to lie with her Master and Mates, who were all extensively educated by years afloat in how hard a vessel could be driven.

Stability is a general term which is applied to different facets of the ship's ability. Initial stability may, in a small vessel, apply to how she feels when you step on board. Sail carrying stability refers to her ability to hold up the rig which drives her, and ought to peak when she has her deck edge down, which is about as hard as most vessels should be driven. Incidentally this stability may be more a matter of hull form, than ballast. Lastly comes the ability of the vessel to right herself from a knockdown. In previous times the vessel would be rigged so that, if all else failed, she could be progressively dismasted by the crew using hatchets on the rope lanyards of her deadeyes. Now that we have full height pole metal masts, and wire rigging, for economic reasons, the ship has to be self-righting from the knockdown without such assistance. It is axiomatic in all sailing ship stability assessments that on some occasion she may meet such a combination of winds and seas that she may be laid over on her beam ends.

These days, therefore, a good sufficiency of stability has to be intrinsic in the design, and to involve self-righting from very high angles of heel. The easiest way for the modern designer is to get the stability range from a deep ballast keel, and then to soften the bilges to get an acceptable rolling motion. This results in the somewhat yachtlike form, which is now often seen, and which would, of course, be less suitable for the no-longer carried bulk cargoes. Load lines are now largely allocated to suit the load condition calculations in the stability booklet, rather than attempting any assessment of the practical cargo carrying capacities (see chapter 14 and annex 14A).

2. Wetted surface

Light air performance is—taken apart from the rig—principally a function of the wetted surface area of the hull. The yacht form has generally a greater wetted surface area for a given volume than the bulkier looking forms of the past, and it is fortunate that the auxiliary engine has largely relieved the designer of this consideration.

3. Hull speed

The shape of the ship has to be planned in relating to its anticipated performance, which—unlike a power ship—is strictly related to windspeed. Two important expressions are:

- Ship speed to root length ratio. $V/\sqrt{L}$
  V = ship speed in knots (knots)
  L = waterline length in feet (ft)
- Theoretical maximum hull speed. $1.4\sqrt{L}$ (knots)

A ship of 100 ft in 30 knots of wind is therefore likely to sail at 10 knots ($V/\sqrt{L}$ of 1), and to have a maximum speed of 14 knots. A ship of 1,000 ft sailing in the same waters, in the same 30 knot wind might sail a little faster—say, 12 knots ($V/\sqrt{L}$ of 0.4)—but despite her theoretical maximum speed of 44 knots, she would need some 150 knots of wind to achieve it, and it would be a brave captain who set the canvas in such conditions.

The potential wave formations in working wind conditions therefore vary heavily with size. A small sailing ship, if sailed competitively, needs a hull form which caters for the wave formation relevant to full theoretical hull speed ($1.4\sqrt{L}$), while the larger vessels are never likely to get anywhere near their maximum speed, and can have aft hull forms which concentrate, for instance, on minimising sea disturbance.

Balance

There is little more complicated in terms of exact calculations than an accurate assessment of the drive centre of the sails of a square rigger. The relative air flows around some 20 or 30 separate soft airfoils over some 300° of headings, and a full range of air speeds, would be daunting enough, without adding in the interface effects of the sea and air boundaries, angles of heel and the variations of sail setting and sheeting. However, experience—in fact over 4,000 years of recorded seafaring—is to hand to give the designer aid. Rigs have always evolved quite slowly and have not really changed materially in recent times.

Hull forms, on the other hand, have had to change, but they are simpler to assess, and the balance between the traditional rig and the variation of hull form can be put forward in the traditional terms of the relationship—or 'lead'—of the geometrical centre of the sails (CE), over the geometrical centre of the underwater profile (CLR). The main principle affecting this balance assessment lies in the actual relative speeds which are achievable by ships of different sizes. A small 64 ft ship will often meet winds of three (or more) times the $\sqrt{L}$ relative speed comparison figure. By contrast, a ship of 256 ft will comparatively rarely meet winds of three times her relative speed figure (16 knots). The first example may need a 'lead' of 15 per cent her waterline length, the second inevitably working at a slower relative speed, will require a lead of 10 per cent (or less), while the largest square riggers often went to sea with little or no lead at all.

Balance in a square rigger is a different matter to that of the ordinary yacht. The former may have seven or eight 'stacks' of sails to the two of the yacht. The change in CE of each individual sail in the yacht can often conveniently balance that of the hull, leaving her relatively well balanced throughout her normal sailing speed range, without any action by her Captain. In the square rigger the effect is reduced to a third—or even a quarter—due to the number of sails, and their relative size. Balance therefore is a matter for the Master to achieve by his sail setting, and the rule of thumb must be that the harder it blows the more he has to reduce his canvas aft (see chapter 4).

The extensive system of jibs and intermast staysails to be seen (especially in the clipper ships), are there to aid the Master in balancing the rig, as much as they are there to reduce rolling.

4. Rolling

The roll motion of a sailing ship is particularly important. It has to be slow and soft enough to allow the crew to work aloft, and not unduly to inhibit the flow of wind across the sails and through the slots. The sheer inertia of the weight of the rig aloft is helpful, but the key to the rolling motion lies in the form of the waterplane at and around the load waterline.

The form of the hull should also include some positive roll damping. This is usually achieved by designing the hull to lift bodily at the extent of each roll. This basically varies the rolling period, and prevents a build up of synchronous rolling.

If done with discretion, some of the roll energy itself can be absorbed in the conversion from kinetic to potential, and back again. Synchronous rolling can also be excited by sea conditions when the rate of encounter of the wave pattern matches, is the same as (or is close to) the rolling period of the ship. The designer can often plan the form to avoid synchronous conditions, which may be common for the ships planned areas of use. However, if these conditions are encountered, it would be normal for the Master to make a course alteration to vary the rate of encounter.

5. Pitching

The rate of pitching has to be considered much as that for rolling, and the hull form planned to avoid any build up in pitch energy. This is usually achieved by a markedly different form of fore-body to aft-body, such that the pitching periods are different, thus preventing any build up of pitch. Vessels with apparently similar ends will normally be designed to have their pitch centres aft of amidships, which effectively provides the necessary differences in fore-body to aft-body. However, some of the largest ships were of such size, that they were unlikely ever to meet a sea which would match their natural pitching period.

6. Dryness

Dryness at sea is a virtue much appreciated by those on board. Essentially two separate conditions have to be considered by the designer:

First: Green Water—which is a matter of great concern as the sheer weight of it will do damage, and the sheer quantity of it be a potential danger to the ship and her crew.

Second: Light Spray—accelerated by the hull form or fittings and flung high enough for the wind to catch and throw back, this is very irritating and dampening, but only a danger to the ship in terms of the condition of the crew on deck.

A sailing ship can be seriously in danger if green water comes on board forward, or aft. The bow has to be buoyant enough to hold up the drive of a lofty rig, when driven hard into a head sea, and not be driven clear under. Seas coming on board from aft—the notorious and much feared pooping—are usually the result of the waterflow accelerating around the curved after-body (such as a counter stern immersed in a wave head) actually sucking the stern down. If the vessel is required to be driven at speed in such conditions, then the stern must be shaped to avoid these inturning curves. An example of this form is the traditional more boxy looking sterns associated with fast frigates and packet boats.

Light spray is generated at the spray roots at the interface of hull and surface water, principally around the bow. It is accelerated—often to quite high speeds—by the pitching energy of the hull itself, and it is this induced power which sends it flying high where it can be caught by the wind and flung back at the ship. The designer in putting some control on pitching can aid the picture, but the shape of the hull around the waterline at the bow is equally important in the initial generation of the spray. Once started it squirts up the hull surface in a slick at right angles to the spray root, which in a bluff bow can actually point forward. It can then be directed and even accelerated upwards by the hull detail, or it can be dissipated and turned more harmlessly away. The important factor is to get this spray to leave the hull surface at as low an angle as possible.

7. Steering

A ship is steered by her own hull put athwart the line of advance by the trim tab effect of her rudder. The shape of the forefoot can therefore be very important in determining steering characteristics, an effect as important as the shape of the after-body in her straight running. A clipper ship therefore could operate with what we would now consider a tiny rudder, whereas our more contemporary cutaway yacht forms need a larger one. Steering astern functions in a similar manner and there must be some suitable vertical aft hull surface to turn her. This is essential if the ship is to be manoeuvred successfully without the benefit of her sails, such as in harbour (see chapter 5).

Nowadays a single auxiliary engine usually results in a propeller in an aperture immediately ahead of or even cut from the rudder. When the vessel is on the wind there will usually be a cross-flow through the aperture, which may result in the stalling of that part of the rudder blade immediately in line with it. The situation falls somewhat fortuitously into a scenario where the subsequent reduced rudder ability occurs with the ship on the wind, where the rig can be balanced, and the rudder area required to handle her is reduced. The stalling ceases as she frees off, and the full area is available for course keeping downwind. Usually the only discernible symptom of a partially stalled rudder in a vessel of any size is a certain hardness on the helm.

8. Manoeuvrability

A modern sailing vessel has to be fairly nimble under both sail and power. Under sail this can be enhanced by a discreet concentration of the underwater profile just aft of amidships, and this also assists when under power. Twin engines are increasingly seen. It is not always appreciated that the total blade area of twin propellers need only add up to that of the single propeller, and with certain reservations they can greatly enhance harbour manoeuvring. A sailing ship rig has really quite extraordinary windage, and even with considerable power on board the ship has still to be berthed and manoeuvred as a primarily wind influenced craft. Bow thrusters can help, but need to be sized against the windage of the rig, rather than the size of the ship (see chapter 5).

Construction

Modern construction methods vary principally from the classic past in the development of the homogeneous skin. No longer do we need to build from a hundred sheets of steel held together with individual rivets, nor do we need to assemble thousands of items of wood and hold them in place with nails and bolts. The advantages of the principal current construction materials can be summarised as follows:

1. Steel

Steel is more or less the standard construction material for ships of any size, because of its economy and the worldwide availability of the necessary craft skills for repair and maintenance. For smaller ships it had a poor reputation for rusting and longevity, but this is no longer the case for good quality construction. Welded steel allied with the new paint systems can produce a good looking hull with a considerable life. The general style of construction varies little from the traditional, except that fore and aft framing is increasingly used in place of vertical framing.

2. Aluminium

The advantages of a lightweight hull are not usually applicable to modern sailing ships, although aluminium deckhouses can assist in obtaining the necessary stability characteristics. Commercial sail is no longer considered for the fast transport of low-volume high-value cargoes and so there is little to attract the sailing ship owner towards the lighter (and faster) constructions, if they cost more—like aluminium. Square rigged sail training ships also usually have little to gain, as they tend to be very cost conscious, rather than looking for maximum speed performance.

3. Cunifer

Cunifer can be welded to steel and has the prime advantage of freedom from fouling. Only the additional first cost prevents its use for bottom plating. However, many sail training ships now operate all year round, so the freedom from the need to slip or dock at regular intervals may swing the balance towards cunifer plating.

4. Traditional wood

Traditional wood construction is very labour intensive which, of itself, is enough to rule out the building of new ships in this manner in Europe. Maintenance is specialised and also labour intensive and the necessary craft skills are already in decline in Europe. Years of use show that it is in itself a thoroughly satisfactory method of building ships, but it is possible that it has been overtaken technically by the new constructions.

5. Cold moulded wood

The laminated wood hull using modern adhesive and protective resin systems is a very good method of construction resulting in the economic use of the timber, a lightweight engineered product, and a seam-free skin, with an overall good appearance. However, it is somewhat labour intensive, and therefore generally expensive.

6. Glass reinforced plastics (GRP)

This is a constructional method seen at its best for multiple production, which unfortunately is not yet part of the square rigger scene. Various one-off processes of construction using GRP over core materials are available, but in square rigger sizes these are uncommon.

7. Fibre reinforced plastics

The whole field of alternatives to GRP as reinforcement is developing fast, and may be a factor in future construction. However, the main thrust is towards ever more lightweight, rather than reduced costings.

Chapter 2
MODERN EQUIPMENT

THE MAJORITY of sail training vessels are custom built, or specially converted, although there are a few museum vessels that run limited programmes. Vessels vary from the very traditional to ultra modern 'super yachts' such as *De Eendracht*. This division is not entirely clear cut, because to some extent all vessels compromise on tradition, by at the very least using auxiliary engines and electric lights. It is somewhat fashionable for big ship and traditional seaman to sneer at the 'yachting' industry, but it is the dramatic growth of that industry, over the last 30 years, that is directly responsible for producing a large enough market to support the development of specialist sensors and equipment for sailing vessels. For example, foul weather clothing has improved out of all recognition, and the availability of small high quality navigation aids—such as Decca Navigator and radar—is now quite outstanding.

This chapter is divided into annexes which look at various aspects of modern materials and equipment. It is not a bible, but a series of notes designed to whet the appetite, and encourage thought. The philosophy behind it is that neither new, nor old, is necessarily the best to use; instead each case has to be evaluated on its merits.

Annex 2A
RADAR AND ELECTRONIC INSTRUMENTS

RADAR AND ELECTRONICS is a very rapidly developing area that is largely beyond the scope of this book. The following notes are purely concerned with practical aspects that are unlikely to be covered in professional courses, or textbooks. However it must be stressed that as a rule this is not an area for economies, as much of this equipment has an impact on ship safety. Specialist advice needs to be sought, the best possible gear purchased, and it must be professionally installed. The rich man who donates a now surplus (obsolete) radar from his motor-yacht is indulging in showmanship, rather than real generosity, and saddles the ship with a set that is not only aging and less than ideal, but also lacks the very valuable makers' guarantee (and goodwill) that comes with a direct new buy.

Radar arcs

Any minor blind arcs caused by the rigging are largely nullified in practice by the fact that trainee helmsmen zig-zag, rather than hold a steady course.

Radar aerial height

The gain in theoretical range is proportional to the square root of the aerial height, so the problems of mounting the scanner up the mast need to be balanced against the actual range in which you are interested. Coach roof mounted sets do achieve quite respectable ranges, even in normal conditions, and, unlike the higher aerials, this performance can be dramatically boosted by surface evaporation ducts (super refraction). Regardless of the aerial height it is vital to ensure that it has sufficient vertical beam width to cope with your maximum sustained angle of heel.

Radar—daylight displays

The advent of the colour daylight display (raster scan) is a great boon in sail training vessels. Colour is not just a gimmick, but provides extra information and is much easier to interpret. It is also far and away better for instruction than the old fluorescent tubes with their daylight hoods and monochrome displays. The TV format is readily appreciated by today's youth, whereas anything else is greeted with incredulity.

Radar echoing area

In an era when scientists are working hard to reduce radar signatures of military aircraft and ships, the square rigger is a splendid example of what might be termed reverse stealth technology! The geometry of the masts and yards of a square rigger forms a very large radar reflector—provided that they are metal, not wood—which is very comforting in reduced visibility, compared to fore-and-aft yachts, who have to rely on pathetically small radar reflectors.

Instruments—digital vs analogue

Digital processing is an advance, but digital numeric displays are not as good as analogue for enabling you to appreciate rates of change, or angular displacement. Ergonomic studies have clearly established this in cockpit design for aircraft, and the shipping world should take note. The 'glass cockpit' in fact largely emulates analogue displays.

The worst example of this seduction by technology occurred when one sail training vessel upgraded from a magnetic compass to a gyro (generously donated). However the system selected provided as its read-out the numerical spot heading in degrees and **tenths** of a degree. Whilst this was doubtless terrific for a VLCC on autopilot, and for feeding heading information to ARPA, it was totally unsuitable for any sailing ship helmsman—let alone a trainee.

Eagle—foremast mounted radar—note the good protective cage for the aerial, and anti-chafe baggywrinkle. This is a relatively long aerial compared to those in most civilian sailing vessels, and it will thus have a larger vertical beam width, and a narrower horizontal beamwidth—both of which are highly desirable characteristics (Rosemary Mudie).

SAILS

By Andrew Cassell (Ratsey & Lapthorn Ltd)

Sail cloths

- **Canvas/flax:** This cloth was used from the 17th century, being produced in two versions. The best quality *(Royal Navy)* was a dark grey colour, and was woven with long fibres spun from the flax plant. The lesser quality *(Merchant Navy)* was a lighter grey with a brownish lint, and was woven from shorter flax yarns, sometimes mixed with other fibres such as jute. Both were produced in a range of nine UK weights ranging from 35 ounces per sq yd for 00 cloth, to 21 ounces per sq yd for 7 cloth (UK weight). *Royal Navy* cloth was closely woven, as it was essential that it had the minimum of stretch and porosity, in order to give warships the best possible windward performance (the weather gauge was vital in battle). *Merchant Navy* cloth was not woven to this high standard, so as to make it cheaper for the commercial market. Both suffered from the same disadvantages:

  a) While soft and easy to handle when dry, when wet they shrink and become stiff and heavy to handle, making stowing the sails more difficult.

  b) If left soaked with salt water over a long period then mildew grows on the fibres (not the cloth).

- **Linen:** This was a flax fibre cloth (mainly produced in Ireland) using yarns that had been laid out to be bleached by the sun. It was still prone to mildew, but the yarns were much finer, so it was popular for light weather sails.

- **Cotton:** Widely used, there were two basic grades of cotton for yachts. The first was *Brown Egyptian,* which had a brownish tinge, and was woven from cotton grown on the Nile delta, with the best coming from the upper reaches in the Sudan. The other type was *American,* which was white in colour, but less strong. In addition India produced similar, but inferior, white cotton. All cotton was superior to flax for yachts, because it did not stretch so much, and it could be proofed against mildew.

- **Jute/Hessian:** Only used in the poorer countries (for fishing boats etc) these materials have similar characteristics to flax, and shrink when wet, but they are far inferior in strength.

- **Nylon (Polyamide):** This material was tried because of its great strength, but in practice its elasticity has restricted it to balloon jibs and spinnakers, where stretch is not so important.

- **Rayon:** This material proved totally unsuitable because it tended to dissolve after prolonged contact with seawater. There was a famous occasion when the 12 metre *Evaine* used a rayon genoa, only to have part of the foot fall off, just before the end of the race!

- **Polyester:** Originally developed by ICI in the 1940s, polyester has been marketed under various trades names *(Terylene, Dacron, Tergol, Tetron)* and is the most widely used modern sail material. It is unsurpassed as a strong, long lasting, low stretch material, although it is susceptible to damage by sunlight in the form of Ultraviolet (UV) radiation. However modern techniques are now providing the fibres with a UV barrier, which greatly improves the life of the cloth. Dying problems have also been overcome with high pressure jigs, and to a lesser extent pad-thermosal transfer, and most colours can now be achieved (with high UV resistance). Coloured sails do fade with exposure to sunlight, and when not in use all polyester sails (dyed or undyed) should where possible be protected from UV radiation (see below).

- **Duradon:** This was developed as a mixture of cotton and nylon, and was intended to take the place of flax as it was much less expensive to produce. In particular it was

Lord Nelson *(right)—mizzen topgallant staysail—roller furling staysails or headsails also always leave one portion exposed to UV, and need a protective strip along the leech and foot (MAX)*

Alexander von Humboldt *(below) —main course yard—the double mainstay is a Germanischer Lloyd requirement. Note that stowing the sails on the yard ensures that the destructive UV radiation always falls upon the same part of the sail, hence the need for sacrificial panels on the aft side of the head of the sail (MAX).*

much touted for square sails, because of its supposed handling qualities. However its achilles heel is that that it needs a white preservative to reduce its porosity (it does not shrink up like flax), and this soon wears off in use (particularly in sails that are regularly folded or rolled up). Thereafter the sails tend to stretch a lot compared to flax.

- **Kevlar/Mylar:** These more exotic materials are not economically viable for sail training vessels, though when laminated to polyester they make good racing sails, because of their low stretch factor.

Chafe protection

All sails are subject to chafe and to alleviate this the sailmaker should reinforce the sail where he considers that this will occur, or where he has been advised that this occurs (dialogue does pay dividends). On-board chafe can be reduced by padding obvious areas on the rigging, and in square riggers this is traditionally done by using *baggywrinkle*. This is an effective system, but it is very inclined to pick up dirt (and grit), which both reduces its effectiveness, and badly marks the sails. Canvas covered padding, and/or soft clear plastic tubing, does the job rather better.

Sail care

All sails (synthetic or natural fibre) should be washed out at regular intervals with fresh water to prevent the salt water crystallizing on the cloth, thus causing mildew in the long term. For synthetic sails this crystalline salt can also destroy the smooth finish applied in manufacture, and make them less efficient.

Sails and UV

Square sails should have a sacrificial panel on the reverse of the head to absorb the UV while the sail is at its most vulnerable, in the stowed position on top of the yard. These panels are easier (and cheaper) to replace than an entire sail, and although they will not be totally effective they do contain the problem, and limit the UV damage. When in harbour in sunnier climes the life of fore and aft sails can be greatly increased by if they are provided with a (loose) sail cover, or totally unbent and bagged up. Roller furling headsails do present a particular problem, but like square sails they can be fitted with suitable sacrificial panels—in this case along the leech and foot.

Sail repairs

For synthetic sails, running repairs are best initiated using white sticky-backed nylon ripstop tape. This lightweight tape is 1.5 ounce nylon, backed with adhesive, and with light sails it alone can be used to repair small tears until final stitching (hand or machine). The surface of the cloth must be dry, and free of grease and grime. Heavier cloth weights (over 10 ounce) will require stitching immediately, and the patches can be held in place, to prevent puckering, by using a double-sided tape. Bolt rope, head, or clew repairs also generally require immediate stitching, but seams can sometimes be held with the ripstop nylon tape as a temporary measure.

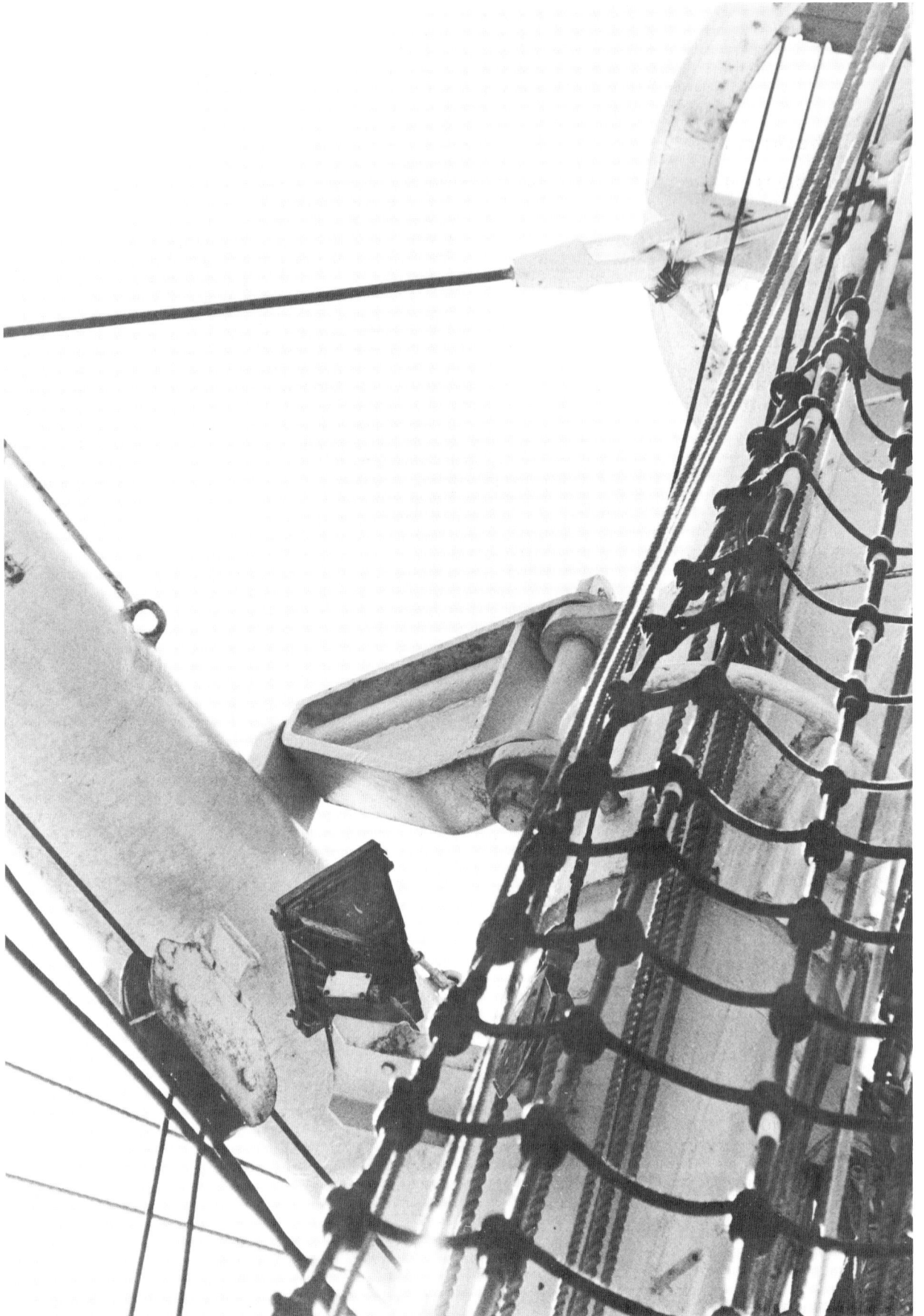

Dar Mlodziezy—*course yard—a heavy duty single point crane is used on this fixed yard—not very elegant, but undoubtedly as strong as a T34 tank. Note the patent terminal fitting on the 1×19 SWR stay (MAX).*

RUNNING AND STANDING RIGGING

Synthetic rope

GOOD QUALITY natural fibre rope is now too expensive to use on board, and in general everyone now uses synthetic (man-made fibre) rope of some variety.

Although for pure strength it is possible to economise and use a smaller diameter with synthetic rope, it is normal to continue to use the similar diameters to those required for natural fibre. This because the rope size needs to be sufficient to provide a good grip for heaving and hauling, and any gain in SWL is incidental.

Terylene (polyester) and Nylon (polyamide) tend to be the most commonly used in running rigging. If the money was available, then Kevlar (aramid core, polyester sheath) could be excellent for halliards, because of its minimal stretch and greater SWL, provided that sharp deflections were avoided.

When using synthetic rope it is essential that its breaking strain does not exceed that of its associated mast, or deck, fitting(s). If a fitting goes before the rope, then the stretch factor will fire the debris back down the line like a missile, and possibly cause casualties. An unpleasant example of this was in the early 1970s in the Dartmouth training ship (admittedly a power vessel), when an eyeplate gave way while using a bull rope to range the cable on the forecastle. It is definitely something to beware of in many older vessels that were designed and built before the arrival of synthetics.

Braidline

In synthetic rope braidline handles better than traditional three-stranded hawser laid rope, and is slightly stronger. In particular some varieties of (cheaper) hawser laid synthetic rope are very prone to crow's footing.

Coding lines

Purists may disagree, but when cadet crews are on board for a very short time (a fortnight or less), it is a good idea to ease their learning curve by using rope of different lay, size, or colour (or a combination), to highlight the different groupings along the pin rails, and fife rails, both visually, and by texture. For example:

- **Buntlines** —braided with blue rogue yarns.
- **Clewlines** —white hawser laid.
- **Downhauls** —red braided.
- **Sheets** —braided with yellow rogue yarns.
- **Braces** —white hawser laid.
- **Halliards** —plain braided.

Stoppers in synthetic rope

The traditional rope stopper (a half-hitch against the lay, followed by wrapping round with the lay) is all too often done wrong, even by those who should know better. However with synthetic fibre rope there is an almost foolproof alternative, which is known in *Eagle* as a 'Chinese' stopper:

- Make up the stopper with two tails. These tails are then simply crossed over alternately above, and below, the rope under tension, thus locking it most efficiently.

Chain sheets

The use of short-link chain for the sheets of the upper square sails used to be standard, mainly because of its resistance to chafe. However, well-lubricated flexible SWR

Eagle *(right)—Chinese stopper—a USCG cadet demonstrates the correct use of a double tailed 'Chinese' stopper on a polyester braidline davit fall (Morin Scott).*

Lord Nelson *(below)—main topsail yard from aft—the large stainless steel single arm crane pushes the yard well forward of the mast for the best bracing angle. Note the special shamrock block underneath the yard to provide a central feed for clewlines and sheets (MAX).*

(FSWR) now has its proponents, who argue that size-for-size FSWR is stronger than wrought iron chain, comparable to high-tensile steel, and very much lighter than either. For sail training vessels (with young mixed crews) the weight factor is significant, as it makes the task of stowing the sails more a test of skill, than strength. In practice wire has worked well in many of the smaller vessels (under 45 m), and has been used in much larger, but those who use chain will probably be loath to change from a well-proven system—and who can say that they are wrong.

1×19 steel wire rope (SWR)

For stays which carry the luff of a headsail (or inter-mast staysail), 1×19 stainless SWR is much more long lasting and less likely to strand than conventional SWR.

Standing rigging—stainless vs galvanised

Stainless SWR (SSWR) is expensive, and not as strong as galvanised SWR, therefore except for the case above it is both cheaper and more practical to use galvanised. There is a particularly good variety of galvanised 7×7/7×19 SWR (Norselay) that is impregnated with heavy duty UV inhibited black plastic (polypropylene). Its unusual 'herringbone' construction (opposite helix in adjacent strands) enables the plastic to form a solid composite section, which is immune to peeling, and protects the wire against capillary action, should the outer cover be cut or abraded. Its corrosion resistance is at least comparable to stainless steel, but for only half the price.

Rod rigging

Solid rod rigging for major shrouds or stays is expensive, and highly susceptible to impact damage. However, it is very suitable for futtock shrouds, where the shrouds and ratlines each side can be made up in rod/bar as one complete unit, as is done in a number of vessels.

Terminals

The traditional seized eyes are excellent for standing rigging, but Talurit has proved entirely adequate (at least in vessels up to 62 m), and it is certainly equal, if not superior, to any basic eyesplice.

There are patent swaged, and swage-less, terminals for 1×19 SWR, which are very effective and simple to use; it would be counter-productive not to use them for this type of SWR.

Rigging screws

Stainless steel rigging screws are an expensive luxury that few can justify, and for the majority galvanised is more than adequate. For ease of maintenance the open-bodied type is recommended instead of closed-body. Titanium is available, but that is completely over the top for a utilitarian sail training vessel.

Shackles, eyebolts and eyeplates

On the principle of 'spend to save,' it is strongly recommended that maximum use is made of stainless steel for shackles, eyebolts and eyeplates, as it will cut down on the maintenance effort (and bill), and greatly enhance the appearance of the vessel.

Mousing shackles

Nylon Insulok ties are a quick and efficient substitute for wire when mousing a shackle pin, and they have the advantage that they will do not chafe on the sails. In addition, use of brightly coloured Insulok ties makes checking the integrity of mousings a very quick and easy task. However, wire must still be used for mousing open hooks, or other rigging fittings.

Tufnol blocks

Some people get heartburn when they see other than wooden blocks. Certainly good quality wooden blocks look beautiful, and have an extremely long life, provided that they are properly maintained. Tufnol (bonded resin fibre) is neither as pretty, nor as robust, but the initial cost is lower, and it needs less maintenance. In some new construction the blocks may be wooden, but they are extremely shoddy, and do not measure up to traditional standards. In such cases Tufnol would be better in every respect.

Special customised blocks

Much is available 'off the shelf,' but some items still need to be custom-made, such as: bulwark course sheet swivel blocks, and central shamrock blocks on yards for clewlines and sheets. Although they are expensive, they greatly improve efficiency, and are **not** luxuries.

Sheaves

If a sheave jams through corrosion, then the line running over it is liable to be damaged. Consequently aluminium sheaves, though still very common, are being overtaken in popularity by new hard-wearing plastic alloys such as Delrin, and Ertalyte.

Marlingspike seamanship

To restore the reader's faith in tradition: a palm and needle sailmaker's whipping still remains the most efficient method of finishing a rope's end, and all the basic skills of knots and splices continue to be important. However, today the seaman must also be proficient in techniques—such as braidline splicing—that are definitely modern. Young cadets are intelligent, and questioning; consequently all teaching needs to be justified as relevant, not just historic. For example, it is somewhat irrelevant to cover the survey and repair of canvas sails, but omit synthetic sails.

The British RYA competent crew syllabus—though not designed for square riggers—in fact provides for a very adequate basic grounding in this subject.

THE RIGGING SCHEDULE

A PROPERLY laid-out up-to-date rigging schedule is invaluable in the long term. Reliance on memory and experience is fine in its way, but loss of the 'expert' through ill health, or another reason, can be crippling during a major refit, if there is no source document to fall back on.

The example included here shows how the rigging schedule can include critical information. In this case the section headed 'Theoretical Loading' shows the safety factor that the designer has allowed, which is why SWR and rope diameters should not be reduced below that specified, although thinner (and cheaper) SWR and rope would cope with normal operations. This safety factor is further highlighted for the eyeplate, under 'Other Equipment.' The specification of special rope is also significant, because of the potential problems involved in surging synthetic fibre rope.

This document can be further improved if it is interleaved with blank pages for the Captain to insert notes on operating techniques and requirements, and also to record the reasons behind any improvements or alterations to the original rig.

The majority of vessels have a regular turnover of personnel, and this document may be the only record of the logic behind the rigging requirements. Departing from the original can be a mistake, as Thor Heyerdahl found out in the first of his papyrus raft reconstructions, when a critical stay was left out, because the experts thought that it was only decorative. That raft fell apart, but for the next raft they followed the ancient illustrations, and it worked. The boatswain on one of the older square riggers was much wiser, and when asked the reason for an oddity in the rig he said: 'I cannot think why it was done that way, but I feel that there must be a good reason for it, and I do not wish to change it merely to re-learn that reason.'

RUNNING RIGGING SCHEDULE

VESSEL	35m AUXILIARY SAILING SHIP

RIG ITEM	Main Sheet	NO.	One Pair

THEORETICAL LOADING Max. 5 tons (shock on gybe)
Handling ¼ ton estimated maximum.

WEIGHTS Total 24.02 Kg
VCG 7.4 from Datum
LCG 14.9 A from Midwaterline

WIRE

Material	Size	Type	Length
Four wire spans under boom. 2 Metres each. 16mm SSWR one end spliced into ring. Other end shackled to boom. (This may be replaced by SS fitting (see note))			

BLOCKS

	Size	Type	No.
1.	4	364	two
2.	4	368	two

All supplied with shackles.

SHACKLES

	Size	Type	No.
1.	19mm	748L	Four

ROPE

	Material	Size	Type (Special*)	Length
1. Port	Terylene	21mm	Braided	44 Metres
2. Starboard	Terylene	21mm	Braided	44 Metres
			Total	88 Metres

*Special = Marlow surging rope.

SHEAVES

1.	None

OTHER EQUIPMENT Sparmaker to supply suitable eyes for spans or fitting at boom end as decided. Shipyard to supply & fit suitable eyeplate capable of withstanding shock load of 5 tons. Two belaying pins, cleats or kevels.

NOTES Wire span system may be replaced by boom end fitting following discussion between designer, spar maker & rigging consultant.

DIAGRAM

ONE SIDE ONLY SHOWN IN SKETCH

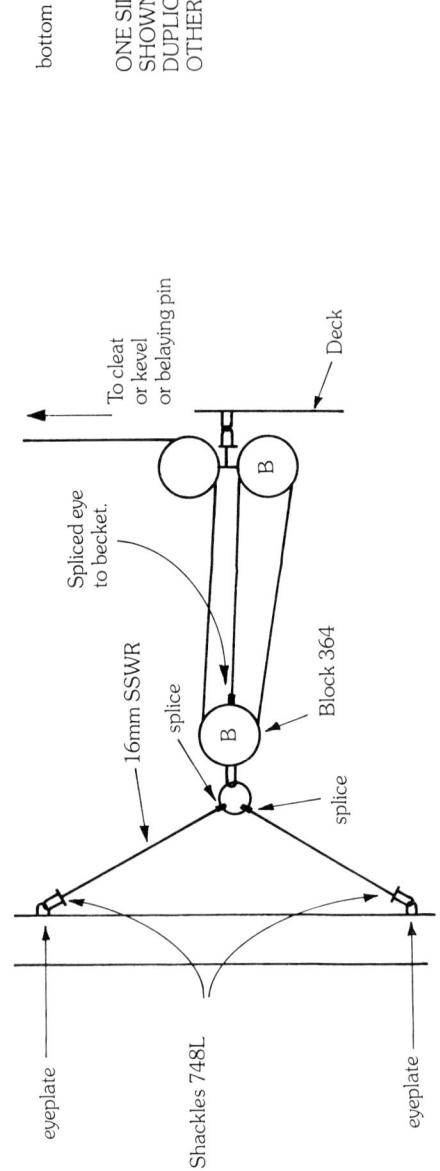

ONE SIDE ONLY SHOWN IN SKETCH DUPLICATE FOR OTHER SIDE

MASTS AND SPARS

Aluminium masts and spars

ANODISED ALUMINIUM is undoubtedly ugly, but painted aluminium is very pleasant. Indeed, in some vessels so fitted visitors are surprised to find out that the spars are not wood. Aluminium is strong, easy to maintain, and can be repaired, or strengthened, by the insertion of a sleeve, and the operation of a simple pop-riveter. Its light weight also makes it a good alternative to steel for the upper spars of larger vessels.

Corrosion in aluminium spars

In all metallic spars, corrosion is a source of concern, but with a reputable sparmaker, good design and construction can greatly reduce the problem. Dissimilar metals (stainless steel fittings) obviously need proper insulation; however, more fundamental is the selection of the correct grade of material for manufacture. Low-quality aluminium (6063) may appear to be cheaper, but any such saving is very short term, because cheap is also low grade and the chance of material failure is thus increased, and early re-sparring becomes a near certainty. It can thus be seen that quality control and selecting the correct high grade of aluminium (probably 6061) is at least as important as good spar design; however, all too often this area is sadly neglected.

Non-anodised extrusions are not protected internally, but this is not a severe problem, because bare aluminium develops a protective oxidised layer on its surface that inhibits further corrosion. This is a great advantage compared to steel-built masts and spars.

The most critical area for mast corrosion is at the heel, because any weakening here could cause the whole mast to drop under compression load, with a potentially disastrous loss of tension in the rig.

Painted spars are normally externally coated with two-part polyurethane, but because of the self-protective qualities of aluminium any damage (scratch, etc.) that goes down to bare metal is largely cosmetic. However, when repairs are done to the paintwork, self-etching primer is essential as the base. Much quiet amusement can be had in harbour watching people diligently putting standard primer on aluminium (or galvanised steel), unaware that it is totally useless.

Fatigue and stress

Fatique and stress are much misunderstood, but for the mariner the actual cause of any crack is somewhat academic. In general, stress can be countered by ensuring that load-bearing points are strengthened, and that standing rigging fittings have sufficient articulation. Fatigue is much more complex, and insidious, and all structures that oscillate (as a spar does) are subject to fatigue cracks. Solutions are not easy; for example, inserting a doubler to cover a known 'weak spot' often merely transmits that problem somewhere else in the structure.

Inspecting for cracks, or fractures, is initially done visually, with the aid of a magnifying glass, and a knife to scrape away paintwork. Full investigation of possible cracks requires non-destructive testing techniques, such as dye penetrant.

Pole masts

A large number of modern vessels have departed from traditional practice and selected pole mast rigs, rather than traditional sectional fidded masts. Considering that combined lower/topmasts were common in the latter days of sail, the use of pole masts is more a case of evolution than revolution, and those who serve in these vessels have a justifiable problem in understanding why the romantics are so upset.

Producing a mast in a single aluminium extrusion is very much cheaper, and lighter, than producing it in separate sections with doublings, and these pole masts have become

Royalist—spanker track—stainless steel track and slides is used for the spanker gaff jaws (as seen here), and for the boom gooseneck fitting, and the luff of the sail (Author).

Royalist—fore topsail yard crane and hoisting arrangement—note stainless steel track, and twin arm crane (Author).

a well-established design feature in modern vessels of all sizes. However, sometimes there are special reasons for having fidded masts—for example, *Gorch Foch* and her sisters were designed with masts in two sections so that they could lower topgallantmasts to pass under the Kiel canal bridges.

Mast exhausts

Using the mast may seem to be an easy answer to the problem of dealing with diesel exhaust fumes in a sailing vessel, but it it not quite that simple:

- The inside of the mast be properly lined, or the sulphur content of the fumes will cause severe corrosion.
- The length of exhaust piping needs to be carefully calculated to avoid back pressure problems.
- The exhaust cannot be totally removed from the disturbed air envelope of the rig—even if it exits at the mast head—consequently there will always be some soot distribution.

Mast track v. parrals/hoops

Hoisting yards can either be set up in parral bands, or in (stainless) steel track mounted on the front of the mast. Track systems were commonplace at in the latter days of sail, and the introduction of stainless steel only serves to reduce the maintenance load still further. Compared to parral mounting, a track and car system also has the advantage of allowing the yard to be fitted with a crane/truss of sufficient size to push it well clear of the mast and shrouds, thus enabling a better bracing angle to be achieved, and improving windward performance.

Likewise, the luff of the spanker and the spanker topsail can be in mast hoops, or set in slides in track, and the gaff jaws can be replaced by a slide. These fore-and-aft sails set better with luff track (particularly the spanker topsail), and no training value is lost for all this gain.

Wooden bowsprits

In order to make the bowsprit sacrificial (absorbing impact) in the event of a collision, or berthing incident, vessels with otherwise metal (steel/ aluminium) masts and spars often retain a wooden bowsprit. Experience with berthing incidents in smaller vessels (every vessel has them at some time), indicates that this is a sensible move that minimises the damage to the vessel, and saves costs because wood is relatively cheap, and quick to repair. Clean, unblemished, wood should be used, preferably Oregon pine, or pitch pine.

Lord Nelson—*main topgallant yard from aft—the roller furling line and fairleads can be seen on the aft side of the yard, and the gypsy at the yard arm. Double jackstays and footropes are still fitted for working aloft for maintenance, and for furling if the mechanical system fails. Note that the royal sheet is wire, and passes through the yard via an internal sheave (MAX).*

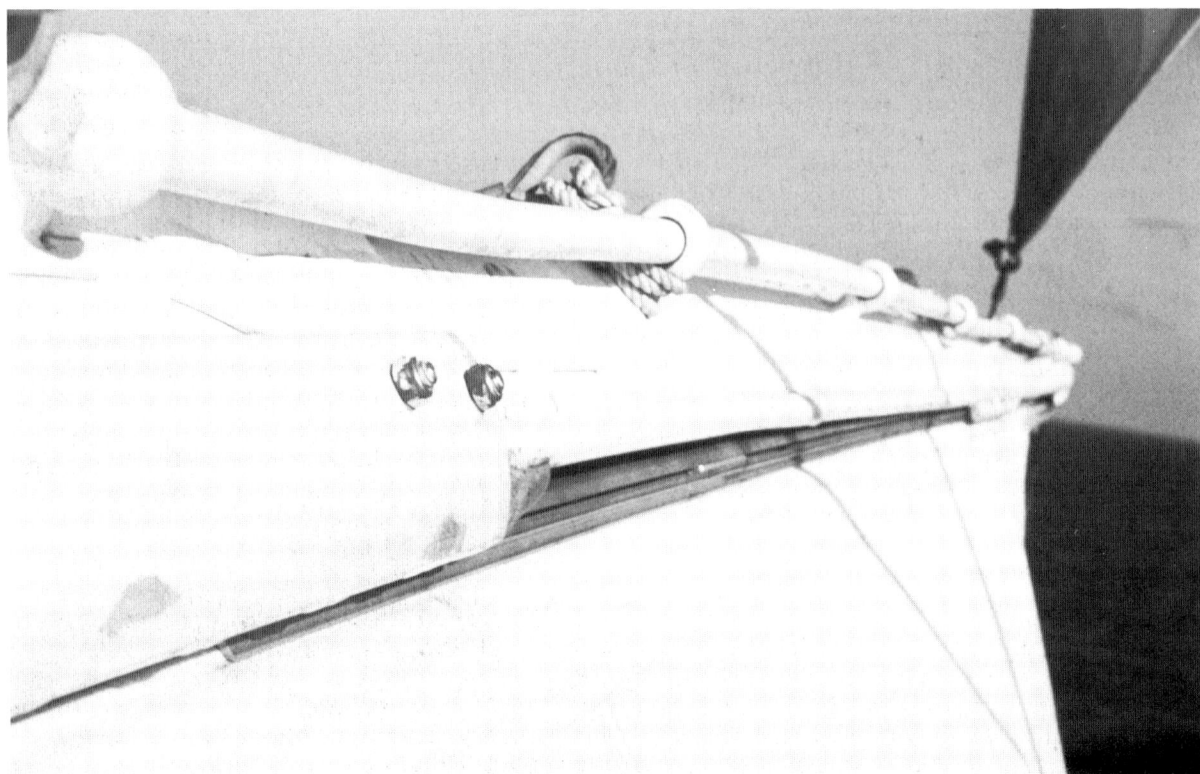

Lord Nelson—*main topgallant yard from for'rard—note the 'letterbox' slot for the sail in the fore side of the yard, and the access plate at the centre (MAX).*

Annex 2F

SQUARE SAIL ROLLER FURLING GEAR

THIS SECTION covers a brief technical description for the modern manual roller furling system, while its handling characteristics are covered later in Annex 4C.

Development background

The modern roller reefing and furling system for square sails appeared with *Lord Nelson* in 1985, and employed the technology that had been developed for in-mast mainsail furling systems for yachts. The idea of roller reefing and furling has been around for many years, but this system represented a step change from the original that dated from the late 19th century and was so extensively used in the French Grand Banks schooners.

These early systems required a parent yard underneath which (or on the fore side of which) was slung a 'roller' yard, connected at the yard arms and also supported by crutches towards the centre. This was an untidy arrangement, which increased both topweight and furled windage.

The new system

In the new system, a much smaller roller is contained within the metal shell of a parent yard, thus not only reducing topweight and windage, but also allowing a much cleaner and more pleasing appearance. An added bonus is that the furled sail is almost entirely protected from UV.

The basic arrangements for crossing the yard are identical to those for a conventional yard, and it still has braces, lifts, and a halliard (for hoisting yards). However, the sail only has sheets attached, as the need for clewlines, buntlines and leechlines is obviated by the (roller) furling line and its associated gear.

The yard

The parent yard is made out of aluminium, and after initial manufacture it has to be specially machined internally in order to remove any blemishes or imperfection that might otherwise cause chafe as the sail was being furled (rolled up), or set (rolled out). Externally the yard is similar to normal yards except for two major modifications:

- **The slot.** A specially strengthened letterbox slot is made along the forward face, through which the sail is set and furled. The extra strengthening is required because the flexing of the yard would otherwise tend to open up the slot, allowing the roller to pop out partially, only to jam when the tension was removed on furling. The slot is much longer than the head of the sail, because it has to accommodate the entire length of the foot, which cannot be 'pleated' like a conventional square sail. When the sail is set the gap between the head earring and the actual yard arm is very striking, even compared to those vessels which have a deep reefing conventional topsail.

- **Roller furling gear.** The back of the starboard half of the yard has all the attachments for the roller furling gear. This consists of two single cheek sheaves in a fiddle arrangement at the centre; two equally spaced stainless-steel double eye fairleads; and an external drive unit at the yard arm.

Furling line

The roller furling gear is controlled by the furling line. This is an endless loop of braided polyester line which leads up from a deck block at the fife rail to one of the cheek sheaves, through the fairleads, round the gypsy on the external drive unit, and then back down to the deck, via the fairleads and the other cheek sheave. Depending upon which direction that the furling line is pulled, then the sail is either set or furled. In the event of the line parting, or a problem with the external drive gypsy, then a winch handle can be inserted into the drive unit at the yardarm to operate directly the bevel gearing inside the yard.

Cheek Sheaves

Roller Furling Drive Gypsy

YARD Ⓢ Ⓢ

Fairleads

Endless braided rope

Fife Rail Belaying Pin

Deck Block

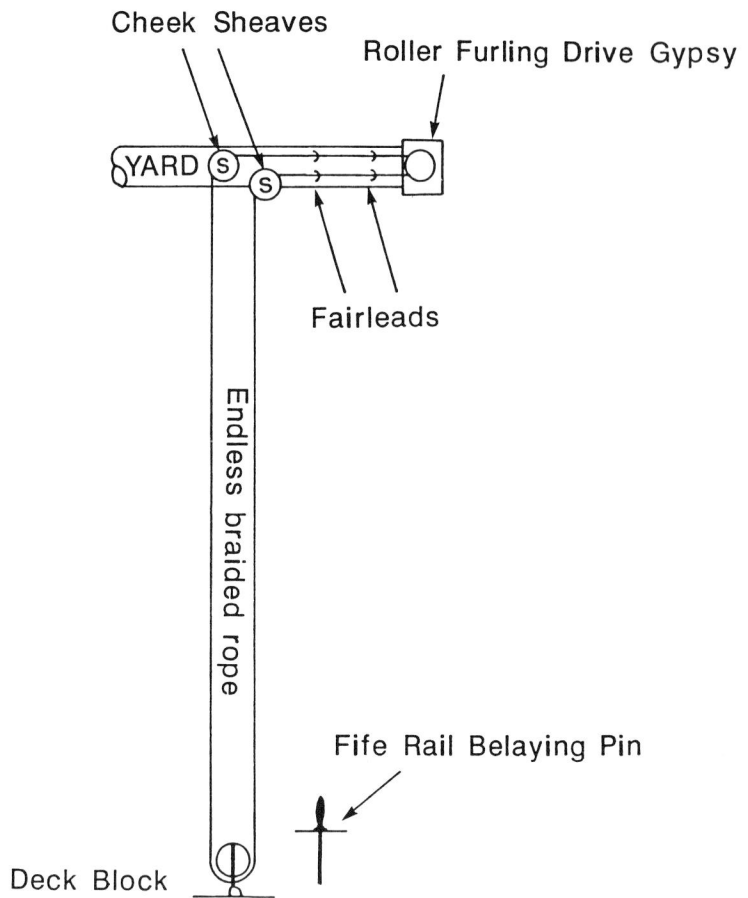

*Figure 2F-1: Roller furling gear layout—-
simplified view (starboard side of yard
from aft).*

The roller or foil

The sail is attached to a roller (rotating foil), which is a section of high-tensile stainless-steel tubing, with a spot welded pressed stainless-steel groove designed to accept the bolt rope at the head of the sail. The roller is driven via a bevel gear box inside the yard, and the direction of rotation is controlled by the external drive gypsy.

In later and more complex models, this reversing drive has been altered so that the gears only engage for the furling operation, and for setting the drive merely trips the ratchet mechanism, thus allowing the roller to freewheel while the sail is hauled out by the action of the sheets (and the halliard—for hoisting yards). This improvement is intended to prevent any possibility of the sail being unwound inside the yard, and consequently jamming.

Attaching the sail

The roller groove has a centre feed with an access plate in the shell plating of the yard. The head is fed into the groove, one side at a time, much as a fore-and-aft sail is fed into a luff groove. Once in position the head earrings are simply lashed to eyes on the circular flange plates at either end of the roller. Again, suitable access is provided at either yardarm.

Maintaining the roller

The roller and the drive mechanism need little maintenance, apart from regular lubrication. However, inspection, or repair, requires removal of the unit, and this is a major evolution which involves both yard arms being dismantled. This operation is only practical once the yard has been struck down and placed ashore and it is normally reserved for refits.

Power operation

The system could easily be adapted for power operation by mounting hydraulic or electric motor drive units at each yardarm (for redundancy), both with emergency manual winch back-up. The motors would have to have overload cut-outs, in order to prevent excessive strain on the sail, and/or the drive units, in the event of a something jamming. A further cut-out switch would be needed at the fully home (furled) position.

Fixed yards

If fixed yards are used, then advantage can be taken of fixed lifts (see Annex 4B). This means that the yard is supported by its lifts at all times, so it remains straighter, thus reducing the stress on the slot.

In addition, vertical clew guide wires could be rigged between the yards, thus ensuring that the clews always roll up into the correct position, and do not flog. The clew should be attached to the guide wire with a roller shackle for best results.

FINISH

START

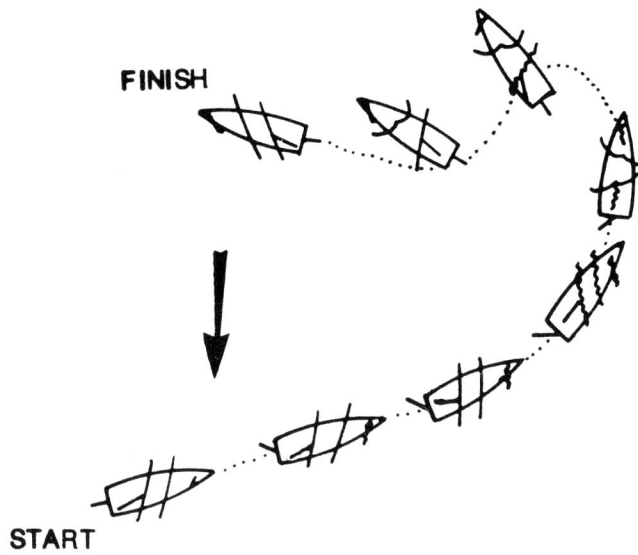

Fig 3-1 Tacking a Brig

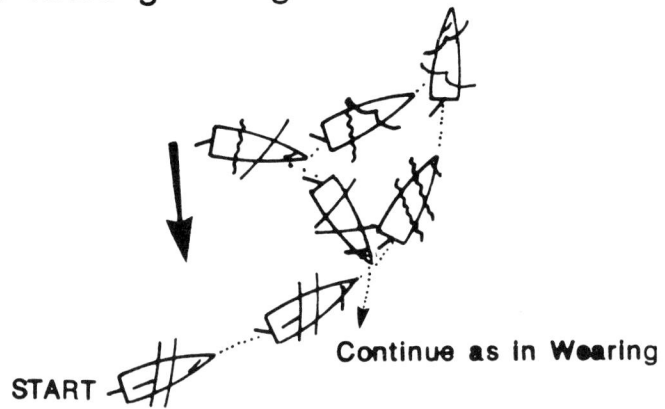

START

Continue as in Wearing

Fig 3-2 Boxhauling a Brig

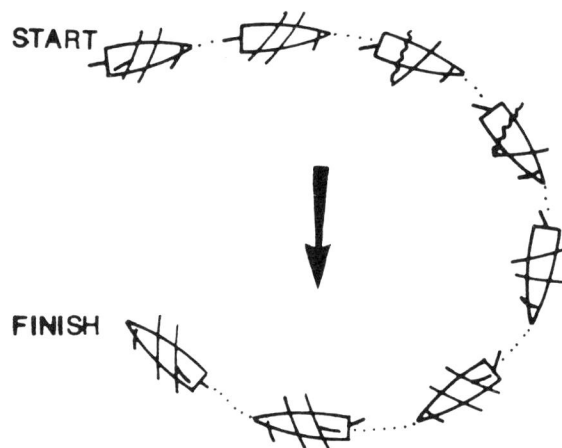

START

FINISH

Fig 3-3 Wearing a Brig

Chapter 3

MANOEUVRING UNDER SAIL

THERE IS A TENDENCY to think of square riggers as slow cumbersome vessels, which were only used on downwind routes. In fact, a square rigger can be highly manoeuvrable; indeed, in Nelson's day, it was generally accepted if two warships of equal force met up, one a schooner, and the other a square rigger, then the schooner did not stand a chance. The ability to back and fill, and boxhaul, gave the square rigger a flexibility in battle that a schooner could not even approach. The schooner might have had slightly better windward performance, but only slightly, and a better crew and better sail trimming in the square rigger could offset that. The British Royal Navy at that time enjoyed a phenomenal kill ratio of better than 13:1 in single ship frigate actions, and there is no doubt that if such innovative officers as Cochrane and Pellew had thought that fore-and-aft rig could improve their results (and prize money), then they would have tried it. Even the Americans did not extend their predilection for schooners to proper warships!

The brig was the most manoeuvrable vessel of all, hence its popularity in both the Navy and the Merchant Service (until manpower became expensive). Yet 20 years ago in Britain square rig was regarded as 'old fashioned,' and the decision to build *Royalist* as a brig was commonly regarded as excessively adventurous, with many confidently predicting that she would have to cut down to a brigantine in order to make her into a practical sail training vessel.

Admittedly, the great wall-sided cargo carriers at the end of the age of sail may have been a little cumbersome, but the large schooners proved infinitely worse, with the largest of them handling so badly as to barely warrant being called a sailing vessel.

Tacking

Tacking can be a long process in a large square rigger, because of the preparations required, but with a modicum of drill a small vessel can be tacked in very short order. Regardless of size, all vessels are different; some being forgiving and others positively cantankerous, but the basics of the manoeuvre remain the same:

a) Maximum boat speed (momentum) is essential at the start of the manoeuvre, so the helmsman is normally ordered to steer 'full and by' to avoid the risk of pinching.

b) Excessive helm should be avoided in tacking, because its braking effect generally greatly exceeds any increase in the rate of turn.

c) Most vessels need to haul the spanker up to windward to improve the turning geometry.

d) The moment to brace the main (or after) yards is critical, and the order 'Mainsail Haul' should be given as these sails stall, so that the yards come round easily and indeed are assisted by the wind. Left too late, they will be aback and the vessel will die in the water before they can be braced round.

e) Small modern vessels—properly handled—will keep way on throughout the manoeuvre, but others may gather a sternboard as the fore yards come aback and will need to reverse their helm at that time.

f) The fore yards are difficult to brace round as they are aback and, depending on the rate of turn achieved, the vessel will sag off to leeward to some extent before they can be braced on to the new tack and the vessel gets sailing again properly.

Tacking—use of headsails

The headsails are often misused when tacking, and this can cause problems. As the helm is put a'lee the headsails should be eased until they stall—but **not** totally let fly. If let fly, then the resultant flogging will not only markedly reduce the working lives of these (expensive) sails, but it will also actually oppose the turning momentum. If, on the other hand, they are correctly handled and merely eased to the stall, then they can be quickly hardened in to back them if the vessel hesitates in stays. Another important reason not to let fly the sheets is that the wildly gyrating sheets and associated blocks can cause injury to the foredeck hands.

In those vessels that are notably unhandy in stays, stalling the headsails may not be adequate, and it may be necessary to drop them at the start of the tack and reset them when through the wind. This is obviously rather inefficient, and in the long term such a vessel would benefit from the attention of a naval architect to recommend alterations to the rig and/or the rudder.

Tacking—handling courses

In commercial vessels it was traditional to ease the workload in tacking by clewing up the courses before bracing the yards, using the order 'Rise up tacks and sheets.' This attitude has carried across to some sail training vessels, but it is often done inefficiently or unnecessarily.

It is inefficient to clew up right at the start of the manoeuvre because it costs valuable drive; indeed, some vessels will not tack if this is done. A better approach is to wait until the square sails are lifting, clew up the mainsail just before the order 'Mainsail Haul,' and the foresail just after. Modern sail training vessels have large crews, and relatively light (synthetic) sails, and can experiment with tacking with the courses set throughout. It is certainly standard practice in most smaller vessels, as indeed it appears to have been in the Royal Navy's training brigs at the turn of the century.

Discussion with German former *Pamir/Passat* officers reveals that these vessels tended to tack with the foresail set. The timing of letting go the fore-tack was regarded as critical, so it was rigged on a slip, and the First Mate had the duty of judging the moment to knock off that slip.

Wearing

Gybing a large schooner can be an exciting business in heavy weather, but wearing a square rigger is almost leisurely. Wearing is normally done when there is either too little wind to tack—or too much—and like a tack it is all levers, and balances:

- First brail up the spanker—because it opposes the turn.
- The helm is then put over (not excessively), and the after yards are braced to weather until they stall, and continually adjusted thus, until they are sharp up in position for the new tack.
- The head yards are squared once the vessel is downwind, and then progressively braced up sharp for the new tack.
- As the stern goes through the wind the headsails, and intermast staysails are 'let draw' on to the new tack, and the spanker is reset.

Wearing—intermast staysails

If a barque has relatively large mizzen staysails, then it is probably necessary to take them in—as well as the spanker—to enable her to wear. This is certainly the case in *Alexander Von Humboldt.*

Boxhauling

An undeserved air of mystique surrounds boxhauling. It is not a complex manoeuvre, but whilst some captains regard it as an elegant little evolution which can be very useful if caught in stays, others would never touch it under any circumstances. Vessels without square rig on the mainmast (such as brigantines and barquentines) do not have the luxury of this option, but for the remainder it is considered that being able to exploit fully the sailing envelope is only seamanlike. In fact, were it to be known by its alternative, and more prosaic, (naval) description, 'wearing short round,' then people would probably be less shy of it.

Boxhauling is only marginally more stressful for the rigging than tacking, and as a rule if the conditions are suitable to tack, then they are suitable to boxhaul. The process starts as for a tack, but as way comes off, and the vessel comes head to wind, the procedure changes:

a) Brail up the spanker.

b) Brace the fore yards round (a'box), and square the main yards.

c) The helm **must** continue to be kept over to lee, in order to allow the vessel to turn on her heel, as her backed square sails make her gather sternway.

d) Once her stern starts to come up into the wind she will lose way, because the after sails will spill their wind, cease to back, and then begin once more to draw properly. At this stage the helm should be reversed (over to weather), and the fore yards braced to fill, so that she can make headway again.

e) Thereafter the manoeuvre is identical to wearing.

> Back in 1777 Hutchinson put the case for boxhauling very succinctly, and today his old English spelling and phraseology adds a certain charm to his words: *Box-Hauling may be proved to be the surest and readiest method to get a ship under command of the helm and sails, to answer many occasions in little room, as well as to ware, and to bring her from one tack to the other, with the least loss of ground to leward, when a ship refuses stays.*

This method of recovering from missing stays (or being caught aback) is most useful in the early days of a cruise, when crew co-ordination (and helming skills) can be less than ideal. In such cases it also has the psychological advantage of retaining the initiative with the command, and keeping the cadets on the go, rather than allowing an ugly pause with the appearance of being defeated.

Clubhauling

Traditional manuals all contain a brief matter-of-fact description of clubhauling, describing it as a method of staying *in extremis*—particularly in an embayed situation. But the issue is clouded, because the oft quoted account of the successful use of this desperate measure (HMS *Magnificent* in 1814), has been shown to have been nothing of the sort. The only other known claimed example is that of an anonymous Naval Captain (obviously bored by peacetime), who is reputed to have clubhauled as a drill, but only after a long period of training and careful preparation.

The flat prose of the Victorian manuals does not make the requirements for this evolution (breaking the cable, rigging a spring to the quarter, etc) any less daunting, particularly for a cadet crew. It also involves voluntarily losing an anchor, which is not something to do lightly. Considering these factors, and the absence of any good example of its use outside the examination hall, this is one case where most would be willing to forego any wish to exploit fully the performance envelope, and accept that one cannot always compete with the days of 'wooden ships and iron men.'

Backing and filling

If manoverboard is exercised under sail using the 'Reach-Tack-Reach' system, then the technique of backing and filling can be used to keep the speed down, prior to heaving-to finally abreast the exercise manoverboard marker. It may equally be useful in other sailing manoeuvres, such as sailing on to a jetty, or even station-keeping in a parade of sail. Such opportunities may be rare—and parades of sail actually under sail are definitely unusual—but the idea is nice.

If you just want to reduce speed, the trick is to brace the main mast until the sails stall, but are not quite aback. In this position drive will be cut, but directional stability will be retained and bracing to fill remains quick and easy. If, on the other hand, the mast is counter-braced fully aback, then admittedly speed is reduced more quickly and more radically, but steerage way is lost, and bracing to fill again takes more time and more effort. If a drastic reduction in speed is required, then it is seldom necessary to brace the main more than a point or so off square, thus putting the sails aback but not fully counter-braced.

Heaving-to

The ability to stop dead in the water by heaving-to is merely an extension of backing and filling, but it is important to determine which mast(s) to back in a particular vessel and how she will lie once stopped. Most will also need some experimentation with their fore-and-aft sail in order to establish the most balanced combination for heaving-to. Brigantines and barquentines have so much fore-and-aft sail that they will probably have to scandalise their gaff headed sails (see chapter 4).

Towing

It may possibly be over a century since a square rigger under sail took another vessel in tow, but there are numerous accounts of it being done, particularly in Nelson's day. At that time using two 'sail tugs' in tandem seems to have been a recognised procedure over long distances, although getting their respective pull balanced out must have required considerable skill.

The essence of the problem is that almost certainly the two vessels will have a dissimilar rate and direction of drift. This does not cause a serious problem when passing the tow, because the approach can be made on a beam reach, slightly upwind of the dismasted vessel. However, once the messenger is across, and you are trying to connect up, the problems really start. Instead of lying comfortably ahead, you may find that, compared to the disabled vessel, you headreach and/or go downwind so much that you start to take up the strain on the tow whilst under bare poles and before you are properly ready. Use of engines may help, but playing around with engines (particularly astern power) is probably unwise with a tow line in the water.

You will definitely want a good spring in the tow as you gather way, and at this stage (at least) motor-sailing is strongly advised for better control. Towing is a complex evolution at the best of times, as scores of naval officers have found out even under 'ideal' exercise conditions. Those who have no practical experience at all in towing are probably ill-advised to start in a sailing vessel.

Chapter 4

SAIL HANDLING AND TRIMMING

Sail balance

THE RANGE of movement of the centre of effort in a square rigger is substantial compared with a modern Bermudan sloop, and as shown in chapter 1 the balancing moment of the pressure in the sails is less than a third of that in the sloop. If sail is reduced evenly top to bottom in response to the changing wind strength, then the rig becomes out of balance with the hull, thus causing excessive weather helm. Instead, sail should be consciously taken in from aft and increased from forward, taking careful note of helm carried, which causes no real speed loss but does make the vessel more responsive and thus much safer. Another factor is that excessive weather helm soon puts the vessel out of limits for the strength and skill of young cadets, thus moving away from the object of sail training.

Balance and the spanker

Reading the memoirs of that famous passage maker Captain James Learmont, it is apparent that the blue water trade had its critics, and he regarded it as the wrong school to learn sailing, because unlike the coastal trade there was not the same need to trim and manoeuvre constantly to make the best of every shift of wind or tide. One specific bad habit that he highlighted has proved remarkably persistent—namely, the belief that windward ability will be improved by sheeting the spanker hard in while close hauled under square sails. This is a serious lack of sailing *nous*, because the average square rigger sails no closer than 60° to the wind, and since trimming is based on the angle of attack—and the spanker can set closer than 60°—instead it should be eased off substantially. Sheeting in harder than necessary both prevents the sail delivering its maximum potential, and also makes the vessel carry much more weather helm. The net result of this misguided over-sheeting is the opposite to that intended, and, far from improving windward performance, it greatly detracts from it.

Partly for this reason some Captains prefer to change down to a trysail in quite low wind strengths, rather than continue with the whole spanker. However, the ultimate in flexibility is to follow the German fashion, and have a split spanker. Interestingly, in 1990 the United States Coast Guard decided that *Eagle* would revert to her original (German) split spanker.

Downwind the problem is different as the standing rigging in a square rigger does not allow the spanker to be set broad 'yacht fashion,' and the spanker tends to generate more weather helm than drive.

Trimming

There is an interesting lack of cross-pollination between those who sail 'state of the art' yachts, and those who sail 'traditional' sailing vessels. The Chief Officer of one large square rigger even admitted that he did not know how to sail a dinghy! Consequently it is not surprising that the dramatic changes in the art of trimming sails and optimising boat performance that have occurred in yachting over the last 50 years have scarcely touched the world of square riggers (or schooners). Those that constantly look to trim the sails to best advantage have enviable dedication, and though sometimes extreme, they have developed techniques that are applicable in all vessels. Sailing a vessel badly is not a virtue, and it therefore behoves us all to understand such things as slot angles and the application of barber haulers. The art of sailing a square rigger did reach very great heights over the centuries, and should continue to develop, instead of being left to ossify and decay.

Shortening sail

1. **Courses:** The argument as to whether to clew up the weather side first, or the lee side first, is as contentious as ever. Often you can avoid the issue by taking them in simultaneously, but in very heavy weather (or short handed) it is better to do one at

a time. Starting with the weather side unloads the system quicker which is important, but it does stress the yard. Weather or lee regardless, it is important to leech up at the same time, and take up the slack on the buntlines, in order to spill the wind, ease the weight on the clewlines, and minimise flogging.

2. **Topsails and above (fixed yards):** Essentially this is the same problem as for the courses, except that everyone agrees that it is the weather side that should be clewed up first.

3. **Topsails and above (hoisting yards):**
   - Off the wind is straightforward, but close hauled the problem is that the yards cannot brace round so far when lowered. therefore it is necessary to sail full and by and adjust the braces, both to allow the yard to lower at all, and to avoid bearing on and overstressing the shrouds.
   - In an emergency (knockdown, etc.) all this is much too slow, as the yards take time to lower and can jam. In addition you cannot furl topsails, topgallants and royals simultaneously. In such conditions you should clew and bunt up with the yard still in the raised position. Since a yard is then unsupported by the lifts it may cockbill, but this is no threat and can easily be dealt with later.

4. **Fore-and-aft sails:**
   - The great advantage of gaff-headed sails is that you can lower them going downwind—albeit less easily and less quickly than is possible with square sails.
   - In yachts gaff sails were superseded by the Bermudan because the latter offered greatly improved performance to windward. This remains true, but there is a need with Bermudan sails to round up when going downwind, prior to knocking in a reef (or dropping the sail), and this can be a problem. Whilst this is containable in yacht-sized vessels, it is not so amusing in a 300-tonner approaching harbour at night in a rising gale.

Sail size and sail handling

As the vessel increases in size it is interesting how quickly fore-and-aft sails approach the handling limits, even with very large crews. In the Royal Navy of the 1860s and '70s some real behemoths were produced, resulting for example in the brig HMS *Temeraire,* whose foresail was 5,100 sq ft (474 sq m). Even that highly-skilled, all-volunteer, long-service Navy, often found such sails unhandleable in a squall. Not even *Sedov* (now the largest square rigger) has a square sail approaching that size; but some of the schooners and topsail schooners do have fore-and-aft sail of that order. In *Esmeralda* in strong weather it was observed that taking each of her large gaff sails required more manpower, and twice as much time, as was needed for her entire suite of supposedly more awkward foremast square sail.

Vary large unboomed triangular sails present the worst handling problems of all fore-and-aft sails—even with trained crews—because not only is the entire load carried on a single sheet, but also the foot of the sail is free to flog, thus further increasing sheet loading.

Scandalising gaff sails

Depowering the rig by scandalising the gaff sails is useful technique on a number of occasions. The procedure is to lower the peak halliard until the gaff is just below the horizontal, and also ease the sheet. As there could be confusion with the routine for lowering a gaff sail, it is important that a positive check is made that the throat halliard is still secured.

Vessels with standing gaffs cannot use this ploy, but there is an alternative method of scandalising, as used in *Regina Maris*, whereby the tack of a loose-footed gaff sail is triced up to reduce the length of the luff by 30-40 per cent. Old illustrations show that this could be combined with lowering the gaff peak. Although little used today, rigging and operating a tricing line is simple, and the hard school of the coasting schooners from which it originated was not one to suffer inefficient systems.

Curiously, scandalising as a technique is only mentioned in yachting handbooks, but not in the seamanship textbooks of the day, possibly because even by the mid-nineteenth century it was restricted to yachts, and the coastal trade, and was thus rather 'below the salt.'

Fanning the yards

In small vessels fanning the yards when close hauled is possibly rather academic as regards performance. However, on long tacks it does help to avoid inadvertent tacking by providing the OOW with advanced warning, by stalling the upper sails first, thus noticeably reducing the angle of heel whilst still inside the recovery envelope.

In larger vessels fanning is demonstrably important for the overall aerodynamics of the square sail stack. This is still judged by eye, and no real experimentation has been done into the optimum angle between the yards.

Square sail 'slot effect'

Associated with the concept of fanning the yards is the vertical slot effect that exists in a square sail stack. It used to be dismissively believed that these gaps between the sails (accentuated by their roach) were a reflection of the low technology and limited aerodynamic understanding of the day, but it is now recognised that they are vitally important to the overall efficiency of the sail stack on each mast, and that in light airs it is possible to have the sails sheeted too flat.

Bracing problems

Given cadet crews and short cruises you must learn to be ever vigilant for the incorrectly tended (or untended) brace/sheet/tack, and stop the action before any damage is done. The standard mistake is a course tack/sheet left fast, particularly when bracing one mast at a time, as in tacking or wearing.

The problem of over-bracing is readily appreciated when the sails are set, but less so when the yards are in their lowered positions. With no sail set, there is no real weight in the braces; consequently the cadets tend to haul taut with extra enthusiasm. Since the yards brace much less far round when lowered they can come hard up on the shrouds just above the tops, with the result that the for'ard topmast and topgallantmast shrouds can be overstressed. The damage tends to manifest itself in the rigging screw and associated bolts/shackles. Therefore it is a good idea to have the boatswain walk round and check away the relevant lee braces a touch before allowing the lines to be coiled down.

Reefing square sails

Despite their large crews, the majority of sail training vessels have split topsails (with no reefs), rather than single (reefing) topsails. This is probably because sail training vessels largely developed from the last days of sail in the 1930s, and that meant commercial sail, as warship sail development had ceased by 1875. Single topsails mean less windage aloft, a sharper bracing angle and consequently a better windward performance. On the debit side, reefing does require both manpower and training. Commercial sail could not afford the manpower, because it had to compete with steam, so the split topsail became standard. Today's crews are large and windward performance is important in maintaining a schedule, whilst achieving the maximum time under sail. For those vessels that only do short cruises there is some argument for splitting, but longer cruises (a fortnight or more) could certainly benefit from the performance improvement and the extra challenge.

Although there is a certain residual fear of reefing, this is mainly based on ignorance. Just as slab reefing has made a come-back in ocean racing, so there are a variety of good systems for quickly reefing square sails, and modern light synthetic sails are a much easier proposition than the old flax sails (particularly when wet). 'Jackline' is very slick, working well even in large vessels, but reefs must be deep, as there is no point in all that labour if the result is a sail that is still too large. Converting a vessel from split to single topsails (as has been done) may require a new yard, because it now needs to be long enough to provide lashing points for the new head earrings and absorb the (greater) length of the reefed head of the sail.

Lord Nelson *(right)—foremast from aft—* this view shows the contrast between the differently rigged sails: roller furling topgallant, topsail with traditional reefing band, and a plain course. The roller furling topgallant has a very 'clean' appearance without the clutter of clewlines and buntlines/slablines. Both topsail and course furl traditionally, but are fitted with slablines going totally round the sail (rather than standard buntlines), and the course is fitted with Jarvis clewlines. Note the dark patch at the head of the topsail indicating the protective sacrificial panel *(MAX)*.

Centurion *(below)—stern view—The tube* for the telescopic brace quarterbooms are shown with the port side extended, and the starboard side housed *(Author)*.

RIGGING AND SAIL HANDLING VARIATIONS

THERE ARE a number of variations to the standard rig, and some less common handling techniques, which are briefly discussed below:

Jarvis clewlines

In order to provide more control over the leech when setting and furling sail, the ubiquitous Captain Jarvis rigged the clewlines to pass through eyes attached to the leech (spacing 3-5 ft), instead of passing directly from the yardarm block to the clew. This clewline system ensures that that the leech is tightly gathered in at the yardarm, makes leechlines superfluous, and simplifies the mechanics of furling the sail. As such it proved popular in the larger commercial vessels (particularly German) because of the marked gain in control over the sail, particularly when furling shorthanded in heavy weather. However, a disadvantage is that it makes it much more difficult to get a good stow, because the leech is held together, thus making the sail bulk out at the yardarms instead of tapering nicely. Sail training vessels are not shorthanded, and whilst smartness is not an overriding factor, in harbour a traditionally rigged vessel will be in a different league, and being a second-class citizen is not enjoyable.

Slablines

Buntlines are attached to the foot of the sail, whereas slablines are led completely around it. These slablines ensure that the sail is very well secured when it is hanging in its gear, particularly when they are combined with Jarvis clewlines mentioned above (as in *Lord Nelson*). However, most vessels have young active crews, and will prefer to continue with traditional buntlines, and save the price of the extra line required for slablines.

Triangular courses

Sometimes humorously referred to as 'Australian raffees,' triangular courses became quite common, particularly for the crossjack. Their main advantage today is the greatly improved arcs of visibility for the OOW, but the reduced handling requirement (only a single sheet) is scarcely a factor in a sail training vessel!

Raffees

A triangular raffee set above the topsail yard was commonplace in the coastal trade, and in the right conditions it can be a good sail. Experience in the STA schooners is that it is relatively easy to set or furl, although taking it in by pulling it down upside down, can seem slightly alien.

'Flying' square sails

When the *Sir Winston Churchill* first came into service (1966) she set both her square sails 'flying' from the deck, instead of having them properly bent on to the yards. The yards were very much an afterthought, and the ugly and unwieldy setting arrangement resulted from the received wisdom of the day in Britain, which held that going aloft at sea was near suicidal, and that square sails were only of use off the wind. In practice, setting and taking in the course was not too bad, but the topsail required a remarkable cats cradle of running rigging. Because of the layout, neither sail set that well to windward, and (unsurprisingly) people were still required aloft to assist the process of setting, or furling. Very sensibly it was all done away with inside the first two years. Both vessels now at least have the sails properly bent on, although they still cannot brace seriously, as that would require a radical and expensive redesign and rerig for the foremast.

The wisdom of the move to traditional rigging was proved when the two sisters (one rigged 'flying') were caught aback by a wind shift, during a demonstration sail-by on the Thames. Whilst the traditionally rigged vessel was able to recover in good order, her sister ended up in a confused tangle.

Centre-furling square sails

Another method of 'avoiding' work aloft was to have the head of the square sail set in track underneath the yard, and control it with inhauls/outhauls, and brails to the mast. This all worked quite well, but to get the sail to set perfectly it proved essential to have someone up the mast for overhauling the brails, and the stow was rather untidy. In the brigantine *Centurion* (1966) overhauling the fore topsail brails required a man aloft at the lower top, armed with a short boathook, which made for an amusing little evolution!

Paracee boom

This is the name for a spare spar rigged to provide a temporary method of booming out the weather clew of the foresail (or the mainsail), stretching the foot, and flattening the sail. It is not difficult to extemporise, indeed *Sørlandet* has for years boomed out her foresail with the aid of such a spar—without, however, devising such an exotic name for it!

This spar has an interesting pedigree, also being known as a *foregirt, fargood,* or a *wooden bowline,* and it is first recorded as being used by the Vikings, who called it a *beitass.*

Bentinck boom

Greatly favoured in the coastal trade, and possibly named after Captain Bentinck (also credited with the triangular course), this was a permanently rigged light spar that was secured to both clews of the foresail with chain. A chain bridle and purchase was used to attach its centre to the deck at the foot of the foremast, and because it needed only two fore-bowlines for control, it was much less manpower intensive than the standard arrangement of tacks and sheets. No vessel is known to use this system today, but it could be resurrected for a replica or museum ship.

Telescopic quarterbooms

This system derived from the short outrigger booms in larger vessels, and was devised for the little brigantine *Centurion* (60 ft/18 m), because she was narrow gutted with comparatively long yards. Consequently there was a problem if the braces were to be led to the deck aft, and kept clear of the standing rigging when braced sharp. This was solved by fitting an aluminium tube across the stern, and having the brace blocks either side fitted to simple spars (quarterbooms) that were slid out and pinned into position at sea, but housed in harbour to make berthing easier.

Winches various

Labour-saving devices are not common in sail training vessels as the main motive power is 'Norwegian steam,' otherwise known as manpower! Every small boat yachtsman knows how easy it is to make a mistake on a winch. However, while for example a riding turn in a small yacht is annoying, but recoverable, the same mistake in a 500-ton schooner could cause serious injury and damage. Winches of all kinds need careful supervision, and can be a mixed blessing in a training vessel.

● **Brace winches:** These were the most famous Jarvis labour saving innovation. They were extensively used in the latter days of sail, and have been retained on board the two large four-masted barques (former German cargo carriers), *Kruzenstern* (ex-*Padua*) and *Sedov* (ex-*Kommodore Johnsen/Magdalene Vinnen*), that are now operated as training vessels. The design is elegantly simple, and could easily be produced for hydraulic rather than manual operation, as indeed has now been done for a commercial project.

- **Halliard and sheet winches.** Jarvis developed halliard winches can be seen in the same vessels as the brace winches. The modern schooner *De Eendracht* has been built with a vast array of hydraulic winches for controlling sheets, and hoisting sails and gaffs. She is a training vessel, but being a sizeable schooner she suffers from the problem of having fore and aft sails that are on the large side (see chapter 4), and she is also intended to be capable of operating in the charter market.

Running coils

Flaking the braces down on deck in running coils is a standard preparation for tacking in the larger vessels, and the only real issue is whether these coils are made in a skewed circle, or a figure of eight pattern. In any case because they are made towards the bitter end, it is important that they are first coiled normally, in order to remove any kinks.

Bidevindstik

The Scandinavian term *bidevindstik* has no English equivalent, but may be freely translated as a 'close hauled hitch.' It is a method of enabling fast tacks, but ensuring that the yards cannot take charge and fetch up bearing hard on the shrouds. Starting on the starboard tack for example, the following procedure would be followed:
- Trim for close hauled.
- The weather (starboard) braces are then made up with the final turn on the pin being a back hitch, which is then stitched in. This is a *bidevindstik.*
- On tacking the starboard braces are only hauled in (not cast off), and then made up on top of the *bidevindstik.*
- The same procedure as before is now followed with the port braces, and all is now ready.
- For the next tack the starboard braces are made ready in running coils, and then at the order they are cast off, except for the *bidevindstik* which terminates the brace. Thus everyone (apart from one man to check for free running) can haul with a will, knowing that the yards cannot take charge, and will be brought up in the correct position by the *bidevindstik.*
- This procedure then can continue for all further tacks, until it is decided that trimming is more important than tacking. The *bidevindstik* can then be taken out and the braces coiled as normal.

Sorlandet—*running coils for braces—cadets flaking down running coils in the braces ready for tacking. Note that the nearest trainee is not wearing shoes. All are in fact wearing safety harnesses, but of the belt type that is unobtrusive, and easily hidden from view undeneath a smock (Author).*

Annex 4B

FIXED AND HOISTING YARD SYSTEMS COMPARED

ALTHOUGH it is traditional that the upper yards in a square rigger are hoisted to set sail, in recent years Polish shipyards have produced a number of vessels to a fixed yard system. This started with the barquentine *Pogoria* (1980), and has continued with a number of other vessels for a variety of flags. Some vessels like *Alexander Von Humboldt* have been built elsewhere, but rigged to this arrangement.

An objective comparison of the two systems has been attempted in the table below. It is based on extensive service with both, and does not attempt to say which is 'better,' because both do work, and the choice of which to use must depend on the customer.

Fixed	Hoisting
Able to rig lifts to truss of yard above (no need to adjust).	Course yard must have adjustable lifts.
Fixed lifts chafe on sails in light airs/calms.	Lifts slack and clear of sail when sail is set and yard hoisted.
Cannot adjust yards in vertical plane.	Yards can be trimmed in vertical plane.
Cannot easily cockbill yards.	Can cockbill course yard easily when required (docking).
Less running rigging— simple to set and furl.	More running rigging—more complex to set and furl (more teamwork required).
Quick to set and furl.	Slightly slower to set. Markedly slower to furl (except in emergency when yard is left hoisted) (see chapter 4).
Difficult to sheet home clews evenly in strong weather.	Clews sheet home easily and evenly (not under tension).
	Hoisting the yard requires considerable manpower on the halliard.
No tracks/parrels to maintain.	
Cannot be reefed—must have split topsails.	Can be reefed— can use single topsails.
Taking in sail—Jarvis clewlines are probably essential to control sail since it is not depowered by lowering the yard.	Taking in sail also lowers the yard, and further lowers the CE, which is important in storm conditions.
	Looks good when everything is harbour stowed—balanced traditional positioning on mast (good for public relations).

HANDLING ROLLER FURLING SQUARE SAILS

ROLLER FURLING is extensively used in yachts for fore-and-aft headsails, but it will always be a rarity in sail training vessels, because its virtues (ease of use, minimal manpower requirement) are to some extent counter to the norm of trying to find employment for a large crew. However, some specialist cases exist, either designed for a partially physically handicapped crew *(Lord Nelson)*, or based on traditional working boats that used such systems *(Belle Poule* and *Etoile)*. Its virtues come into their own in any modern commercial sail design, so powered roller furling was specified for the stillborn 1980s project to build a 176 m five-masted barque (110,000 sq ft/10,280 sq m), and was first fitted to the 116 m four-masted barquentine *Star Flyer* (35,500 sq ft/3,300 sq m) commissioned for the cruise market in 1991.

The section below is based on the arrangements in the barque *Lord Nelson,* because she was the first square rigger—as opposed to a topsail schooner—to be extensively fitted with modern roller furling/reefing square sails. Although design variations do have an effect on handling, that tends to be only in minor detail, so much of the experience gained in her should read across to other vessels and variants. The actual mechanics of her system are described separately at Annex 2F.

Principles

The principles involved in handling roller furling are remarkably similar to those of a conventional square sail, except that there are no clewlines, buntlines, or leechlines to handle, because they are replaced by the furling line. Experience has shown that this system works, and that the roller furling square sails can be set and furled quickly, often with the minimum of crew, without anyone being required to go aloft.

Problem areas

Two common problems have arisen with this manually powered system:

- When the yard is in the lowered position and the sheets are cast off to allow the rest of the sail to be gathered up, then the clews tend to wind in towards the centre of the yard. This makes for a large bundle at the outer ends of the foil, sometimes filling the yard and making the operation stiff (and chafing the sail). Keeping tension on the sheets has been tried, but without real success.

- If furling is attempted when the yard is not square, then the lee clew will not be gathered up properly, and a triangular piece will be left sticking out. This can be largely avoided with experience, by tensioning the sheets as required.

Setting/furling a roller furling square sail

For both operations the principles are the same as for any square sail, and the orders for the halliards, sheets, and braces, remain as normal. However, furling line orders replace those for clewlines, leechlines, and buntlines. The sequence for setting and furling the main topgallant is used here as an illustrative example:

- **Setting.** At the order 'Stand-by to set main topgallant':
  (i) Hands stand by sheets and furling line—two for each sheet, and three for the furling line.
  (ii) Check that those on the furling line are on the correct part of the line, so that when they haul away it unfurls.
  (iii) Slack away the braces as normal, and have a team ready on the halliard.
  (iv) When ready order 'set main topgallant—haul away furling line and sheets.'
  (v) When the sheets are fully home and reported secure, and the sail is fully unfurled, hoist away on the halliard. The furling line **must not** be secured until the yard is fully hoisted.
  (vi) When the yard is fully hoisted the furling line can be 'loosely' secured, much as buntlines are loosely secured for conventional square sails.

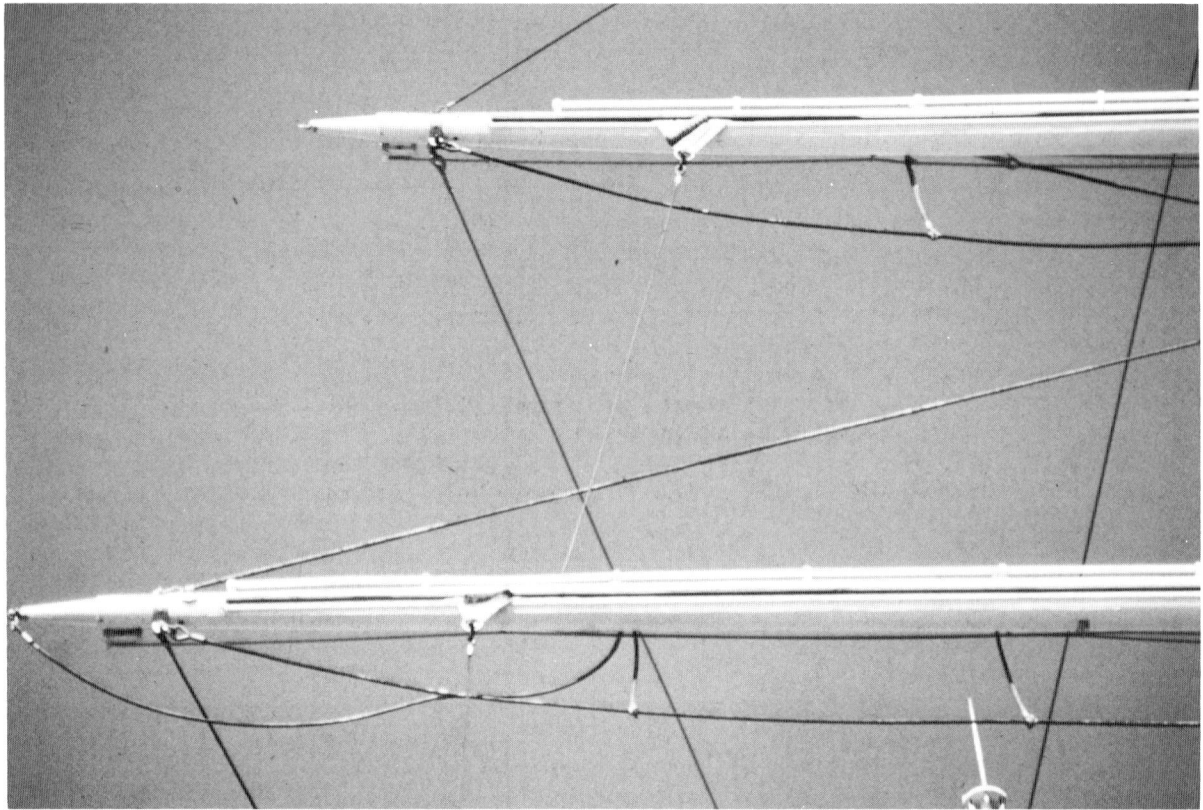

Lord Nelson—*fore topgallant and royal sails roller furled—the clews protrude from the slot very obviously, and well in from the yardarm (MAX).*

Belle Poule—*fore topsail—a good view of the traditional double yard roller furling system as used in the French Grand Banks schooners. Note the considerable remaining windage of spars and sail, compared to the internal roller furling version, or to traditionally furled square sails (Rosemary Mudie).*

- **Furling.** At the order 'Stand-by to furl main topgallant':

(i)   Three hands go to the furling line, and one to the halliard.

(ii) Check that they are ready on the correct part of the furling line. A mistake here is more critical than in setting, because the sail can try to roll up the wrong way, and may jam, or even be damaged.

(iii) If the yard is braced up sharp—ease the lee brace.

(iv) When ready order 'Lower away on the halliard—heave away furling line.' The furling line acts like the clewline in hauling the yard down, while partially furling the sail.

(v)   When the yard is fully down in its lifts, order 'cast off sheets,' and continue hauling on the furling line until the sail is fully rolled up (furled).

(vi) Finally, the furling line is 'loosely' secured (as in setting).

Chapter 5

AUXILIARY POWER AND THE SAILING VESSEL

THE EXTENSIVE and complex windage of a square rigger means that under engines she does not handle as easily as a power-driven vessel of comparable size, particularly in a confined space with a strong wind blowing. In addition the bowsprit is vulnerable, and it can be difficult to judge its position and that of the bobstay relative to obstructions and jetties. This problem is compounded because sail training vessels are very much in the public eye, and their berths are often dictated by their prominence, rather than their ease of approach. Ceremonial can also have its drawbacks, such as the difficulty of passing engine and helm orders over the noise of an exuberant military band playing on the jetty!

Propeller loading

A number of the twin-screw vessels have small high revving propellers, and these can suffer from excessive blade loading when used for manoeuvre, because each prop only has half the blade area that a single-screw vessel would possess. If too much power is applied, the resultant excessive blade loading will result in a breakdown in water flow, and a heavy reduction in power output. The blade area, while sufficient for moving ahead, will prove inadequate if you decide to rush in to your berth with great flair and panache (in true destroyer style), and attempt to pull up with a quick burst of astern. There are occasions when you need to drive in positively, but even then there are limits.

Propeller drag under sail—twin-screw *versus* single-screw

The extra drag created by having twin screws compared to single is not as great as it might seem, because you need the same blade surface area to propel the vessel, whether it is concentrated in one propeller or distributed over two props. The extra drag in a twin-screw vessel therefore comes from having two (smaller) sets of shafts, bosses and brackets, compared to one (larger) set. There is some advantage in hiding a single prop in an aperture, but this is largely negated by the extra deadwood area, and/or the larger rudder area required. Under sail the cross-flow through the aperture causes turbulence over the rudder, and thus greatly reduces its effectiveness. Such vessels often carry excessive weather helm at low speed, and need 'boundary fences' to stabilise the flow over at least part of the rudder in order to improve overall handling and performance.

Propeller drag under sail—shaft trailing

There is some controversy about whether shafts should be locked when under sail, or whether they should be left free to turn. Some people find it difficult to appreciate that a trailing prop still creates similar blade drag to a locked prop. Perhaps it helps to have a helicopter background, for there a major safety factor is the ability to control safely your rate of descent after an engine failure by 'auto-rotating.' The analogy is not exact, but it is nicely illustrative. There may seem to be little to choose between the two apart from the wear and noise, and that varies according to sailing speed and circumstances. However, the principal drag criteria appears to be how much sailing energy is dissipated in turning the propeller(s), and associated shaft(s) and machinery, in their various bearings and glands. Generally those that lock tend to do so on the intuitive grounds that a system should either be providing drive (as designed), or properly secured. In fact, most marine gearboxes rely on an engine-driven pump for lubrication, and therefore need to be modified before trailing, or—alternatively—have a special 'sailing' clutch fitted to allow the shaft to rotate separately.

Motoring to windward

Given the windtrap aloft, your ability to power dead to windward for any distance is severely restricted, and can be a desperate waste of fuel and money. The actual true wind speed limit for any vessel depends on its rig and engine power, and the sea state at the time. If you are faced with a long leg under power it is important that you make every effort to minimise windage, which means bracing the yards and putting a harbour stow on the sails. Above that limiting wind speed (15-20 knots) you will be forced to tack to windward either purely under power (>25° to the wind), or motor-sailing using a combination of power and some fore-and-aft canvas (>40° to the wind).

Motor-sailing

There are many occasions when speed under sail is insufficient, and the extra boost provided by the engines can enable the schedule to be maintained. It is often the case that while the journey could be done under sail alone, there is insufficient time in the programme. In such cases motor-sailing is an economical tactic, as less power is needed for a given speed by using both sail and power, than it would be under power alone, and therefore fuel consumption is kept down. In a lumpy sea left after a blow a vessel tends to be thrown around under sail alone, but the application of power provides much needed momentum, and the resultant increase in speed can be dramatic.

Off the wind it is all gain, but once the wind is for'ard of the beam the increased speed through the water also brings the apparent wind further forward. If already close hauled under square sails the effect of engaging auxiliary power will be to bring the apparent wind too far forward to continue with them. In such a case the fore-and-aft sail should be retained, but shifted to a sharper sheeting position (or **Barber** hauled), to ensure that while the engines provide the majority of the power, the sails contribute valuable additional drive, and significantly dampen the motion. Several vessels have special motor-sailing sheeting positions for their headsails and inter-mast staysails, to ensure that they can hold on to them closer to the apparent wind than they would need to under pure sail.

When motor-sailing leeway is obviously markedly reduced, so the technique is also very useful for weathering a point, clawing out of an anchorage (see chapter 7), or making up round a buoy.

In twin-screw vessels it is good practice in bad weather to run the leeward engine only, as this provides better drive and less cavitation. Steering is also improved, because any weather helm carried puts the rudder into the prop slipstream, thus enhancing its effectiveness.

Windage and manoeuvring in a confined area

Windage has considerable impact on all your manoeuvres in harbours and anchorages; as always there is no substitute for knowing your vessel, but generally the following points should be borne in mind:

1. **Berthing:** Coming alongside under power in any strength of wind the angle of the yards to the wind is still very important. Additionally, since the yards project considerably even when braces hard-up, ideally you should brace the yards to present a 'trailing edge,' when coming alongside another sailing vessel, or an obstructed berth, but this requirement is subordinate to the windage factor. You must consider whether the wind on the yards will slow you down, blow you off/on, etc., and how you can alter, improve or alleviate these effects. In a square rigger your overall windage is such that down-wind berthing is as much to be avoided as down-tide berthing, and when wind and tide are opposed it can be difficult to judge which will be the more important force.

2. **Turning:** Use of the jib can help a lot when turning through the wind. This is particularly important when the wind is from ahead, but just onto the jetty, and you are trying to get off. Conversely, the trysail can be useful when turning into wind, particularly in brigantines or barquentines, which are unbalanced by being square on the foremast only, and thus tend to resist bringing the bow up into any strength of wind.

3. **Sternboard:** Going astern the stern will naturally seek the wind, and even twin-screw vessels do not always steer well astern. To steer well astern a vessel must have either the hull surfaces aft to steer around, or a quite excessive amount of power.

Single rudder—twin screw

Twin-screw sailing vessels are laid out in merchant vessel style with the propellers and rudder out of line. The more manoeuvrable Naval 'twin-screw—twin-rudder' standard is unfortunately only really suitable for power vessels because of the problems of sailing rudders in such an arrangement. In addition to this obvious limitation, the underwater shape of a sailing hull in this configuration creates some interesting shiphandling characteristics that are not present in similarly configured merchant vessels, and are therefore not properly covered in the standard publications such as *Crenshaw:*

1. **Turning with way on:** The novelty here is that with more than 3 knots headway the rudder effect appears to be reversed if you go astern on the inboard screw. This is best explained in terms of deflection of the prop jet. Being out of line the rudder needs to be turned into the flow of an ahead turning propeller to behave normally, whereas going astern on inboard screw causes loss of both the normal flow due to ship speed, and the propeller slipstream. The results are extreme and it can be difficult to accept that with more than about 3 knots headway reversing the *inboard* screw only causes the vessel to cast the 'wrong' way, whereas she turns very well with the *outboard* screw reversed—all of which rather contradicts normal shiphandling practice and experience.

2. **Turning at rest:** Once way is lost, the screws work in the normal sense and you turn logically. In some vessels the screws are well separated for maximum differential force (pressure and suction, and paddlewheel), and it is standard almost to disregard the rudder, leaving it amidships and avoiding nugatory work by the helmsman. In other vessels short bursts on the inboard screw against full rudder works well.

3. **Springing on to a berth:** The textbook method of driving in the stern by going ahead on the fore spring with opposite rudder does not work in all vessels. Instead you may have to use the head rope combined with ahead and astern power and walk the stern in.

Single screw

There are some advantages in having a large and slow revving single propeller, but such an arrangement lacks the inherent flexibility and redundancy of twin screw. Different factors need to be considered, and more space is needed when manoeuvring with single screw:

1. **Turning with way on:** Assuming a right-handed fixed pitch propeller (or a left-handed CPP), in light airs the obvious way to turn short is always to start to starboard. In a strong wind from the starboard beam a port turn would work far better, because when going astern the stern seeks the wind. However this is theory, and in practice things do not always work out so easily!

2. **Turning at rest:** Bursts of ahead power against full rudder work well, but the same factors apply in deciding the best direction as in turning with way on.

Variable pitch propellers (VPP)/Controllable pitch propellers (CPP)

There are advantages to CPP/VPP particularly if they are fully feathering. But they are expensive and complex, reduce the efficiency of the propeller, and they do introduce shiphandling problems of their own. CPP/VPP do not provide a simple alternative to a reversing gearbox, and a seized pitch change mechanism can really ruin your day! For a twin-screw vessel the slight gain in low-speed handling is more than counterbalanced by the loss of the ability to turn at rest—particularly given the very restricted spaces in which one is expected to operate without a tug. By contrast, in a single-screw vessel the argument is much more closely balanced.

Bow-thrusters

Some single-screw vessels are fitted with bow-thrusters. These work best when the vessel is virtually at rest and can be a real asset in turning, pushing the bow on (or off) a jetty, and in controlling heading when going astern. However, in strong winds your windage is such that the power of the bow-thruster may prove inadequate. Despite the propaganda, a bow-thruster is not a panacea, nor is it a substitute for second propeller, and basic shiphandling principles still apply.

Seaboat as a pusher

Using the seaboat to push the bow round, or to push on to a jetty, works well within its obvious limitations, and can be a useful ploy for those vessels without bow-thrusters.

Anchors

For manoeuvring purposes, an anchor can almost be treated as another engine. Obviously it is important to check that there are no underwater obstructions/cables, but otherwise the anchor can be used as follows:

- **Turning:** The procedure for turning is to let go at slow speed (putting the helm over), heaving taut when way is lost, and then turning on it. Since the cable will grow aft throughout there is a possibility of a bad nip on the hawse pipe, which could cause the cable to part under sudden stress. Consequently bursts of power should be avoided.
- **Dredging:** If the anchor is towed at short stay it is termed dredging. This ploy enhances control of the bow, assisting in holding it upwind—or upstream, and can be very useful when berthing in a strong offshore wind.
- **Berthing/Unberthing:** In an onshore wind/stream the anchor should be let go when stopped some distance off and parallel to the berth. The let-go position should be such that when secured alongside the anchor grows abeam with sufficient scope to hold when hauling off. For unberthing, the stern must initially be sprung off, and then held off by using the engines against the cable and a breastrope to prevent the vessel overriding the cable. There are circumstances when a long scope is not practical, and then the inshore anchor should be used.

Coasters regularly use this procedure. However, those trained in the deep-sea trade, or the Navy, though well aware of the theory, will for a variety of reasons tend to have little actual experience, and may need practice before it becomes second nature.

Chapter 6

HEAVY WEATHER

THERE ARE two parts to the problem of heavy weather. The first is how to recognise that such weather is on its way so that preparations can be made, or avoiding action taken; and the second is how to handle the vessel safely and efficiently if caught out. As regards these issues there is a famous report by Admiral Chester W. Nimitz, USN, in which he reflected on the loss of three destroyers (and the damage to many other vessels) in the disastrous 1944 typhoon, and made some fundamental points that have become ever more valid in this increasingly hi-tech age. Although he was writing about power-driven vessels, the essence of what he said is well worth summarising:

● Seamanship is about taking action in ample time. To begrudge action on the grounds that it may prove unnecessary, is to fly in the face of all recorded experience.

● Modern forecasting services in no way relieve the command of the burden of making their own observations and interpretations. Indeed, to ignore the evidence of one's eyes, because the radio report states that there is no storm on its way, is nothing less than culpable negligence.

● We understand a fantastic amount more about the weather than did our forebears, and that means that even when deprived of shoreside support we should still be able to produce a reasonable forecast.

Forecasting

There is no short cut to learning how to interpret the weather, and carry out basic forecasting; on the contrary it is strongly urged that the study of meteorology, and to a lesser extent oceanography, is something that should continue over a career, rather than merely being brushed up for exams. It is also felt that the advent of the modern lightweight weatherfax endorses this position, as it brings such a host of information within reach of the mariner, and enables him to rise above the level of single observer forecasting. Indeed, when you have access to such information the limitations of the BBC forecasts are thrown into sharp relief, and many regard it as an essential item of ship safety equipment.

Heavy weather preparations

The standard preparations of battening down remain most valid and important, particularly in view of the risk of downflooding (see Chapter 14). But when facing exceptional conditions—rather than 'ordinary' gales—it is also necessary to reduce windage aloft. The old textbook reaction at sea (or in harbour) was to strike down upper masts and spars. One reads of a squadron of Victorian battleships housing topgallantmasts in a gale in nine minutes, but such achievements mask a casualty rate that would be totally unacceptable today. Generally apart from the uppermost (and smallest) yards, striking masts or spars with a cadet crew is not a practical proposition at sea. However, windage can be reduced substantially and quickly by unbending all except storm canvas. This unbending should include the spanker, particularly in those vessels where the peak is fixed. In harbour everything possible in the timescale should be sent down on deck, and lashed down with exceptional care.

Upper deck lifelines

Larger vessels will need to rig lifelines across the deck to reduce the chance of people being thrown across the deck and injured. In addition it may also be prudent to rig netting in the vicinity of forecastle and quarterdeck ladders as an added security factor, if swept by a heavy sea.

Running rigging—security

Once you start to dig in and ship solid water, the coils on the bulwarks will be swept off their belaying pins, and trail over the side in a tangle. They are then difficult to recover and ready for use, thus slowing down any handling of the associated sails or spars. In addition, if you have freeing port doors, and the lines wash out the ports, then the tangle of lines will restrict the water flow, until the action of the doors eventually cuts them through, which is even less acceptable. The grills on the freeing ports—now required by the DTp—do reduce the chance of lines being swept through.

The coils of running rigging should therefore be taken clear, and hitched either to the lifelines, or up at shoulder height on the shrouds. This is best done with light line that can be cut in a hurry, but even a gasket hitched coil made up with itself is quicker and easier to handle than one that is trailing over the side.

Heavy weather shiphandling

As late as the 1950s, the standard response to storm conditions was to heave-to under storm canvas, or a lee cloth spread on the shrouds of the aftermost mast. Riding to a sea anchor was also an accepted practice. Modern sails have made a great difference; the racing fraternity have shown how techniques vary greatly with the hull form, and how much a yacht can take (albeit with a highly skilled crew), and the sea anchor is now almost totally out of favour. Square riggers and the larger fore-and-afters have also been affected by these and other developments:

- The most obvious change is that sails in good condition and properly cared for are less likely to blow out at critical moments; however, 'good condition' is the operative word. The accursed accountants (who seem to be everywhere these days) must be made to appreciate that since sails are the primary means of propulsion, it is essential that they are well made and kept in order, rather than being subjected to cheeseparing economies.

- Those vessels with traditional hull forms would probably be overadventurous were they to depart from the old established practice of heaving-to, and they could even consider a sea anchor. However, those vessels with a modern hull form are unlikely to ride easily to a sea anchor, and will not gripe so much.

- Modern vessels with their different hull form, and improved handling and responsiveness, may consider running (scudding) downwind at high speed, taking the seas 20° on the quarter. This idea originated with the legendary Vito Dumas in the 1930s, and is now accepted practice in modern ocean racing. Hydraulic steering as currently installed is unlikely to be adequate for this, as it is too slow. However, the remainder of modern vessels may actually be better off than conventional yachts, because a square topsail is a good deal easier to control than a spinnaker, the rig is more balanced, and if the sheets are freed a touch the topsail will actually generate some lift at the bow, instead of driving her down.

- Towing warps astern to reduce speed is the more common method of running down sea, but it is the shape of the stern that is critical—with the right shape surfing at 25 knots is possible, but with the wrong shape you will be overwhelmed if you run off at hull speed.

Use of oil

The value of oil in storm survival is questioned by many authorities, but equally others would still swear by it. Even if you do believe in it, you still have to decide whether the risk to your vessel justifies the minor pollution. Smaller craft with relatively low freeboards should also consider the danger to life and limb that will result from any of this oil being swept across their decks.

Rounding-up in a storm

If you are running before a storm the time may come that the risk of a broach is such that you need to round-up, and heave-to. Obviously the timing and execution of the turn will require careful judgement, and this is another occasion with the auxiliary engine can be a great boon. The trouble is that initially the rudder will have little flow over it, and therefore its effectiveness is reduced. However, if the engine (centre-line or inner/weather unit) is used for a burst ahead at the start of the turn, then the momentum will be greatly increased at a critical moment. This technique was certainly successfully used in anger by USCG *Eagle* in her 1947 hurricane experience.

Storm sails

A nice shallow lower topsail—or a well-reefed single topsail—is one part of the equation when riding out a storm, and the other is provided by the fore-and-aft storm canvas. These storm jibs and trysails need to be carefully designed not just for their strength, but also for the ease and speed with which they can be set. Setting these sails should be regularly exercised in clement weather, so that the Captain can reassure himself that all the gear is on board and working correctly. This may seem a statement of the blindingly obvious, but it is depressing how often bad weather shows that those responsible have 'assumed' rather than 'checked.'

Scudding

The fore lower topsail (or reefed fore topsail) should be set with the storm jib, and possibly a reefed foresail, combined with some intermast staysails to damp out some of the roll. There is merit in slightly freeing the square sail sheets to provide some lift at the bows. The vessel will roll violently, and only skilled helmsmen should be employed, or the risk of broaching will be excessive. USCG and RNLI power boat experience may be pertinent, because they teach that you must initially avoid countering a broach, because that only stalls the rudder; instead you should let hull speed (and rudder flow) accumulate, before attempting corrective action.

Heaving-to/lying-to

Setting trysail, main topsail (reefed or lower), and storm jib, a ship (or barque) lying to the sea should hold quite comfortably four points off the wind and sea. The topsail is the primary roll damper, and the remainder merely provide balance. The position of the helm will have to be established by experiment, and while some may be able to lash the helm, others will have to attend it constantly. The best position for each vessel must be determined as early as possible in her life, and recorded in the ship's data book.

Smaller two-masted vessels (brigs and brigantines) can obviously carry less canvas, and may have to experiment more to achieve a good balance. In general they would not carry square sails in such weather, and would use a main staysail instead. For reasons of balance a barquentine would have to adopt a similar posture.

In the context of small vessel storm survival, it is worth noting the famous case of the brigantine *Varua*. She started by being hove-to, then lay a'hull, and finally ran off under bare poles, streaming warps to reduce speed from 6-7 knots down to 3-4 knots. So if your vessel is labouring and suffering, it is vital to take heed and think about alternative tactics, because there may something that better suits her.

Storm survival lessons

In addition to the above points there a some important general lessons that have been learnt over the years:

- Broaching-to is not necessarily disastrous, regardless of how violent and frightening it may have been.
- Despite free surface effect, waterlogged vessels have a remarkable record of surviving storm after being abandoned, thus confounding their original crew.
- Engine-operated bilge pumps are invaluable for a prolonged period of pumping in storm conditions.
- In exceptional storms lying a'hull is not a viable tactic.
- Wheels and steering gear are vulnerable to damage.
- Exceptional storms are not textbook events, and survival depends on experience, improvisation, and skill of the highest order. True seaman make their own luck.

Dismasting and jury rigs

Storm damage can include total, or partial dismasting, and at least one sail training vessel has been dismasted in recent years through rigging failure. Others have suffered potentially serious gear failures, but have managed to contain them.

If dismasted you will not be totally divorced from your masts and spars, but will remain attached on the lee side, and possibly some other points. This gear needs to be cut away before it batters a hole in your side—far from easy if you are lying over with the rigging screws for your lee shrouds under water. Hydraulic wire cutters are light and simple to operate, and seem a must for this job. Indeed for 1×19 SWR (see annex 2C) they are probably the **only** practical weapon in this situation. However, while cutting away this gear everything that can be salved will be to your advantage. It may be possible to leave some of it secured by only one wire, to act as a sort of sea anchor, and this could then be recovered later—when conditions have improved. It is unseamanlike to leave any floating wreckage that might endanger, or damage, another vessel.

Jury rigs can work surprisingly well as many accounts testify, though the pictures show that they are seldom pretty. Such spars as can be salvaged can be set up using bulldog grips ('U' must be on the tail), or fist grips (symmetrical—almost foolproof), to make temporary eyes for the standing rigging. Those with pole masts are likely to have rather less to work with than those with fidded masts, but each case will be different. However, these days the temptation will be to accept salvage assistance, in which case only the initial clear-up is necessary.

Chapter 7

COASTAL PASSAGE PLANNING

By Commander David Gay, MBE, MNI, RN (Rtd)

No gentleman ever sails to windward (anon)

THIS CHAPTER concentrates on coastal passage planning, highlighting the following points:
- Windward ability of modern square riggers.
- Weather appreciation (see also Chapter 6).
- Passage planning strategy.
- Embayed situations.
- Anchorages and lee shores.

Three other important factors are not included, as they are covered in detail in other chapters:
- Shipping problems—Chapter 8.
- Traffic separation schemes—Chapter 8.
- Motor-sailing—Chapter 5.

It is targeted at small to medium square riggers (under 500 tons) operating in northern European waters, and does not consider either commercial sail, or fore-and-aft rigged vessels.

The problem of long-distance passages is also left out as this is very adequately covered in the planning publications (see Appendix 2). Effectively the blue water square rigger should stick to the trade routes, where for most of the time the threat of lee shores and embayed situations is remote, and the main concerns are tropical revolving storms, and calms in the doldrums.

Windward ability

Crucial to the ability of any square rigger to make a coastal passage is her windward ability, tacking angle, and best point of sailing. This is clearly dictated by:
- Wind strength and direction.
- Sea state.
- Tidal stream.
- Proximity of land.
- Size and rig of vessel.

For example, compare the tacking angles of the brig *Royalist* (80 tons) and the topsail-schooner *Sir Winston Churchill* (250 tons) in force 4-5 in open waters. One would expect this to vary from some 140° made good between tacks for the brig, to about 120° for the schooner. Clearly this is a major factor in coastal work.

All sailing vessels are happy to be 'tucked up' under a weather shore, where with little sea, and perhaps a fair tide, they can achieve their best speed, both through the water, and over the ground. Square riggers are no exception, and it is very stimulating to sail a brig under full sail and at maximum hull speed, on a coastal passage in these conditions. In contrast, having to beat up the Solent against a foul tide can be very frustrating, although it is good sail drill! In that situation the brig's tacking angle made good would increase from its norm of 140°, to possibly as high as 180°! All of which gives one a great sympathy and respect for our ancestors, and shows that it is futile to fight the combined effects of wind and tide.

In passage planning, therefore, you should strive to avoid tacking into the full force of the tide, and aim to reduce windward work to a minimum, by selection of destination—making full allowance for leeway. It is essential to adjust the sail carried to the prevailing sea state. An under-canvassed square rigger trying to work to windward in a short steep sea will take a very long time to complete its passage.

Weather appreciation

The key factor in all coastal work is good weather forecasting and appreciation. In the case of square rig this becomes a paramount consideration, with emphasis on early and consistent interpretation of weather patterns prior to any passage. As stated in Chapter 6 there is an abundance of information available today to supplement the basic radio shipping forecasts, and every source (even the TV) should be used to build up a picture of the weather situation in the 48 hours before the passage starts.

At sea, a strict forecast discipline needs to be exercised; never miss a forecast— particularly in unsettled conditions. The importance of having recourse to local weather information cannot be overemphasised.

Anticipation of a pronounced 'back' or 'veer' in the wind can result in a fast passage to an attractive port; in contrast a long beat (or motor-sail) will ensue if you get it wrong. This makes it most important not to become too fixated on a particular choice of destination, as this only serves to constrain your ability to plan a sensible passage in line with the prevailing weather pattern.

Finally, you must consider the visiblity, particularly in the spring/early summer and autumn, when morning and evening fog or mist can be experienced in a cross-Channel passage. If one has a choice in such conditions, and the tide serves, then a forenoon or afternoon landfall may be advisable.

Strategy considerations—1

Planning the passages for a typical sail training week in the English Channel is a good means of illustrating the problems involved:

● Let the vessel concerned be a small square rigger, which is equipped with two diesel engines, capable of driving her at 6-7 knots in fair weather.
● The situation is that the wind has just veered to a north-westerly force 4-5, in the wake of a cold front. An unstable westerly weather pattern predominates, and the next depression is due to make itself felt in two to three days time.

The following three questions therefore need to answered by the Master, before he sets sail:

1. **WHEN** should he make passage after the initial crew shakedown?
2. **WHERE** should he aim for? (given the windward ability of his vessel).
3. **WHAT** special considerations must be made in arriving at his best option, and **WHAT** fall-back options does he possess?

Looking at all this, the Master should think along the following lines, and consider:

a) Choosing a course which will lay the port of his choice with minimal tacking, and thus without over-stretching his new crew.
b) Selecting a course to cross the shipping lanes at a reasonable angle. This **must** be as near as practicable a right angle for a traffic separation scheme (see Chapter 8).
c) Avoiding area of high traffic density—if possible.
d) Working the tides. Timings should be such that the tide is used to assist coastal work—for example, the double tide proceeding east past Dover, or being at right angles for the great part of a cross-Channel passage.
e) Having an ETA that is compatible with any locking in requirements, in order to avoid unnecessary time spent at anchor.

In this situation the north westerly force 5-6 makes Cherbourg the most attractive option, when starting from the east Solent. On this route the wind should be on (or abaft) the beam the whole way on the outward leg—provided that the vessel sails in good time. If the departure time was to be delayed over-long, then it is possible that the expected back ahead of the next front could catch you still short of your destination. However, assuming that this does not happen, and you do make it on schedule, then after a short period in harbour you could take advantage of the next wind change, which would be to south east or south. This would allow you to undertake a short hop to Alderney or Guernsey, returning to Britain as it finally veers to the south west.

This may all sound too good to be true, but it really is possible to work the weather pattern to your advantage, and to sail the resultant triangle without the wind ever being for'ard of the beam.

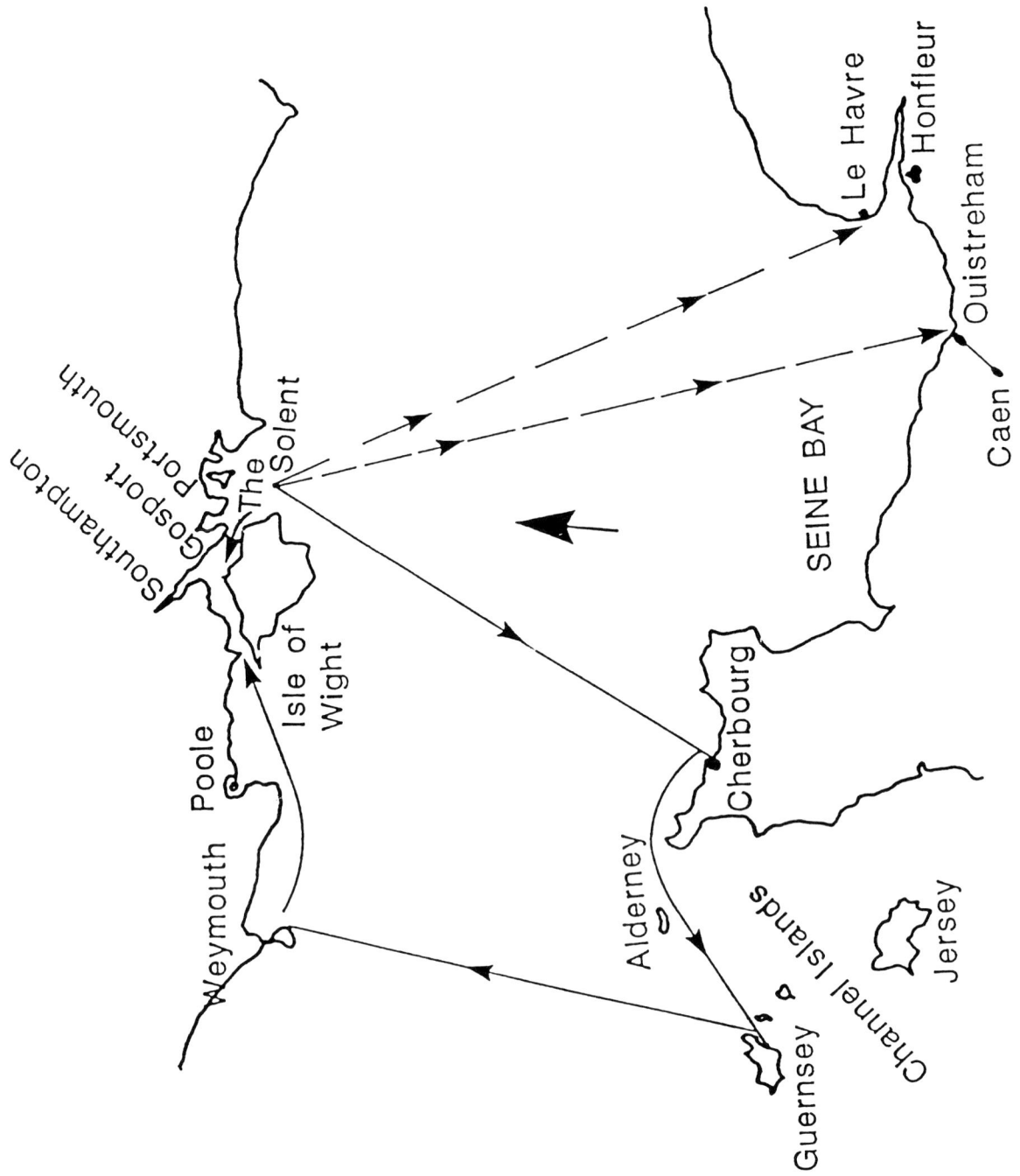

Fig 7-1 Channel Passage Ports

Strategy considerations—2

If, on the other hand, the whole area was dominated by a stable south westerly or north easterly airstream, then there would be a temptation to go for the Seine bay area—Le Havre, Honfleur, Ouistreham, or Caen, according to the exact wind direction.

Even in moderate conditions, fast passages could be expected, both out and back, because sailing with the wind on the beam/quarter is the square rigger's best point of sailing. To sail dead downwind is quite appalling tactics, as it is much better to tack either side of the rhumb line, keeping the wind a good 20° off the stern. The resultant higher boat speed more than compensates for the extra distance sailed, as the results in Tall Ships races have amply demonstrated.

Embaying

The risk of being embayed is often held up as a unique hazard of square rigged sailing. In fact, every prudent seaman will plan his passages to provide an offing, and being embayed is nothing more than the problem of how to extricate oneself from being caught on a lee shore bordered by two headlands. Weatherly vessels can accept lesser margins than unweatherly ones, but there are some desperately unweatherly schooners, so this problem is not the exclusive province of the square rigger. Modern navaids and weather forecasts provide little excuse for putting your vessel in a corner from which it cannot be extricated under sail alone. Since motor-sailing greatly reduces leeway even at reduced power (see Chapter 5), applying the sail alone criteria should provide a good margin for error.

There is a body of opinion that many of the classic cases of disaster through being embayed on passage were attributable to poor navigation compounded by poor diet. A large number of cases happened when making landfall after some weeks at sea, and the diet at sea, until relatively recently, was not only deficient in vitamin C (scurvy), but also lacking in vitamin A. One of the earliest symptoms of vitamin A deficiency is night blindness (nyctalopia), and without radio navaids you could not avoid what you could not see.

Anchorages and lee shores

The problem of being caught and embayed at anchor is a separate and distinct form of being embayed, and being caught at anchor is a serious possibility. For example, *Royalist* was severely embarrassed by this form of embayment very early on in her career, in Studland Bay, when the wind went round from light southerly to north-easterly force 6-8, in short order, and the decision to sail was made far too late. This was a major error of judgement, but weather can be misread, and not even the most competent and conscientious can forestall every eventuality.

Obviously prevention remains better than a cure, and a very careful selection of anchorages is needed, having due regard for the actual and the likely wind directions. The British Admiralty Pilot for the area is always helpful concerning shelter and holding ground. Its advice is based on a historical data base, consequently the Pilot reflects the past experience of sailing vessels, as well as what suits a modern coaster.

Good ground tackle is essential, and a square rigger needs about 50 per cent heavier gear than an equivalent sized fore-and-aft rigged vessel. Boxing the yards restricts the amount that the vessel sails about her anchor, and does help prevent dragging. However, should the anchors fail to hold, and a serious dragging situation develop, then it is normally necessary to weigh and proceed to sea with all dispatch. In these circumstances a powerful windlass is essential for weighing, after which you will need a combination of main engine(s) at full power and steadying fore-and-aft canvas (motor-sailing) (see Chapter 5). These tactics when executed correctly, and in a prompt and positive manner, have proved successful on many occasions.

Conclusion

To sum up the advice in this chapter:

● Watch the weather like a hawk. Use all available sources of information (including calling for special VHF forecasts); make your own observations of the sky; always be sensitive to any backing or veering of the wind.

● You must make your plan according to the weather pattern, and not allow yourself to be seduced by any preconceived plans as to choice of destination.

● Always maintain a good offing in unsettled conditions when in the vicinity of bays, or a lee shore.

● If caught embayed at anchor: carefully brief and supervise the anchor watch; veer cable; make all preparations for clawing off under power and sail.

Chapter 8

THE RULE OF THE ROAD AND
SHIPPING PROBLEMS UNDER SAIL

THE RULE OF THE ROAD is in some ways unsatisfactory as it is really written as if large sailing vessels do not exist. In fact there may be a case for an M Notice to make people aware of the problems and limitations of square riggers. After all their numbers are increasing, and there are quite a number of over 3,000 tons.

The following Rules specifically apply to sailing vessels:

Rule 3c	**General definitions—sailing vessel.**	
	(provided that propelling machinery . . . is not being used).	
Rule 9b	**Narrow channels.**	
	(sailing vessel *shall not impede* safe passage).	
Rule 10d (i)	**Traffic separation schemes.**	
	(sailing vessels *may* use inshore zones).	
Rule 10j	**Traffic separation schemes.**	
	(sailing vessel *shall not impede* safe passage).	
Rule 12	**Collision Rules amongst sailing vessels.**	
	(if in doubt you must assume that the other vessel is on the starboard tack).	
Rule 18a	**Responsibilities between vessels.**	
	(power gives way to sail rule).	
Rule 18b	**Responsibilities between vessels.**	
	(sailing vessel as regards other vessels).	
Rule 25	**Lights and shapes—sailing vessel underway.**	
Rule 35c	**Sound signals for a sailing vessel in restricted visibility.**	
	(morse 'D' signal is shared with five other categories of vessel).	

Further to the above there are other rules or points of interpretation that are pertinent:

Rule 3f   **Not under command (NUC)**
In discussing the circumstances when NUC lights/shapes in **Rule 27a** might be considered appropriate, *Cockcroft and Lameijer* seems to reject adverse weather as adequate grounds, whilst accepting being becalmed. For once this interpretation is flawed:

a) An auxiliary sailing vessel that **chooses** not to use her engine while becalmed, cannot claim the **Rule 18** protection of NUC status—even if she is racing and would incur a penalty for use of engine. The racing fraternity may not like it, but the engine is available, and the yacht racing rules are not a consideration in the International Regulations for Preventing Collisions at Sea.

b) As regards adverse weather, there by contrast the case for NUC would seem to have been underestimated. A sailing vessel (even a large one) is much more at the mercy of wind and sea than a power driven vessel, and can be forced to **heave-to** or **lie a'hull** because she has become impossible to control safely in any other way. Since she has thus been rendered *unable to manoeuvre as required* because of the *exceptional circumstance* of the severe weather this would seem a clear case of being **not under command.**

Rule 8f   **Action to avoid collison**
This rule only came into force in 1989, and is important as it defines the application of the phrase *not to impede* as in **Rules 9b and 10j**, and shows its complementary relationship to the other requirements of the steering and sailing rules.

Rule 10c   **Crossing traffic separation schemes**
The 1987 amendment *on a heading* overcame the previous poor wording, but it is less generally appreciated (despite M 1281) that a sailing vessel

Figure 8-1   Braced Square –
Arc of manoeuvrability without bracing or tacking

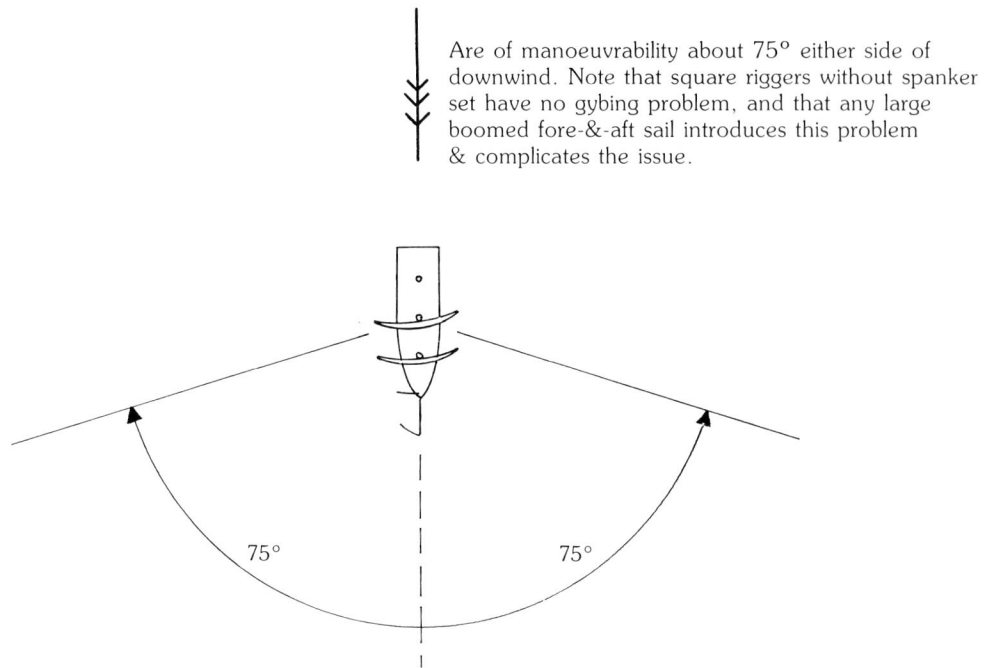

Are of manoeuvrability about 75° either side of downwind. Note that square riggers without spanker set have no gybing problem, and that any large boomed fore-&-aft sail introduces this problem & complicates the issue.

75°   75°

Figure 8-2   Braced up sharp for close hauled – Arc of manoeuvrability without bracing or tacking

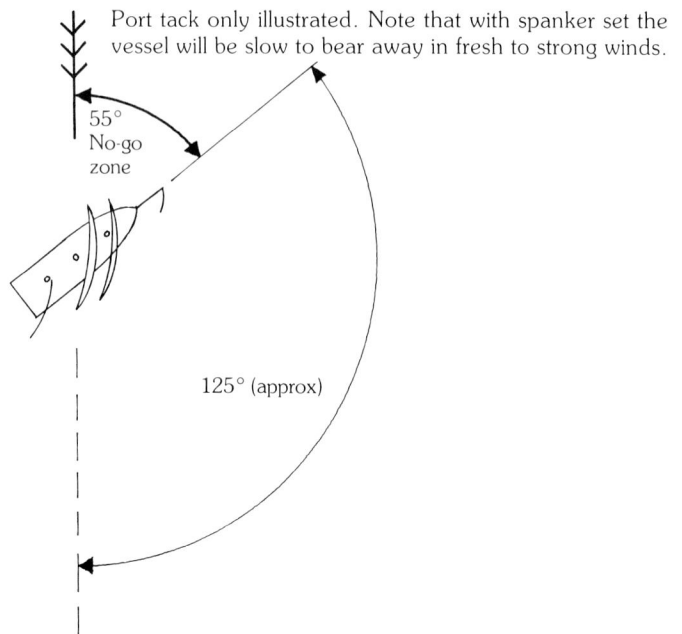

Port tack only illustrated. Note that with spanker set the vessel will be slow to bear away in fresh to strong winds.

55°
No-go
zone

125° (approx)

must use her auxiliary engine when needed to as nearly as practicable achieve the right angle. *Cockcroft and Lameijer* quotes the case of the Dutch schooner *De Eendracht* in August 1986. She was crossing the separation scheme off Terschelling when the correct right angle course was dead down wind, and elected to sail 30° off that heading to avoid the risk of a standing gybe while sailing by the lee. The Court found the OOW guilty of contravening the rules because he did not use the auxiliary engine to assist in achieving the required course.

Square riggers do not have this downwind control problem (none of those large gaffs and runner backstays), but they do have a problem area—into wind. Tacking to windward a square rigger can make no better than 55° to the wind (plus leeway), and she is quite unable to drive directly to windward under power alone in any serious strength of wind. Depending on rig (windage), engine power, and sea state, between 15-20 knots of true wind is enough to stop any square rigger (see Chapter 5), which leaves two choices:

a) Tack to windward under power alone (very slow and >25° to the wind).

b) Motor-sail to windward with the assistance of fore-and-aft canvas (faster but >40° to the wind).

If the wind is near dead foul neither of these will be good enough to come really close to the right angle. Admittedly you could put up a good defence in court for your actions, but it is preferable to avoid that nausea, so it would be sensible to discuss the problem on VHF at the time with the relevant controller.

Rule 12	**Collision Rules amongst sailing vessels** These rules are now written for small fore-and-aft vessels. The fact that port gives way to starboard, regardless of point of sailing, does not take into account the problem of tacking a square rigger.
Rule 13	**Overtaking** Curiously, some people are unaware that if one sailing vessel is approaching another from more than two points abaft her beam, then **Rule 13** over-rides **Rule 12**, because she is an overtaking vessel. Therefore she must keep clear, regardless of the wind direction and the respective tacks of the two vessels.
Rule 17a	**Action by stand-on vessel** The permissive early action allowed under **Rule 17a(ii)** can be a double-edged sword for a sailing vessel because power vessels tend not to appreciate the manoeuvring problems of a square rigger, and therefore seldom alter as early as one might wish. As a result you could find yourself altering almost all the time. However, what most Captains do, is only to take this action for those vessels that for some reason make them wary. Regrettably there seem to be an increasing number of vessels that induce this mild paranoia.
Rule 17c	**Action by stand-on vessel** The statement here that when taking avoiding action under **Rule 17a(ii)** a power driven vessel should 'if the circumstances of the case admit, not alter course to port for a vessel on her own port side' also has a commonsense application to sailing vessels.
Rule 34b	**Manoeuvring and warning signals** Strictly the signals indicating an alteration to port/starboard only apply to power-driven vessels. However, logic indicates that they should also be used by sailing vessels taking avoiding action under **Rules 17a(ii) or 17b**. (Although there was potential for confusion with the sailing vessel fog signals in the previous (1960) Rules, this is not the case in the present (1972) Rules.)

Shipping problems under sail

A square rigger attracts interest at all times and it is sometimes difficult to determine whether a collision situation is developing, or whether the other vessel merely wishes

Figure 8-3 — Rule of the Road Problems between sailing vessels

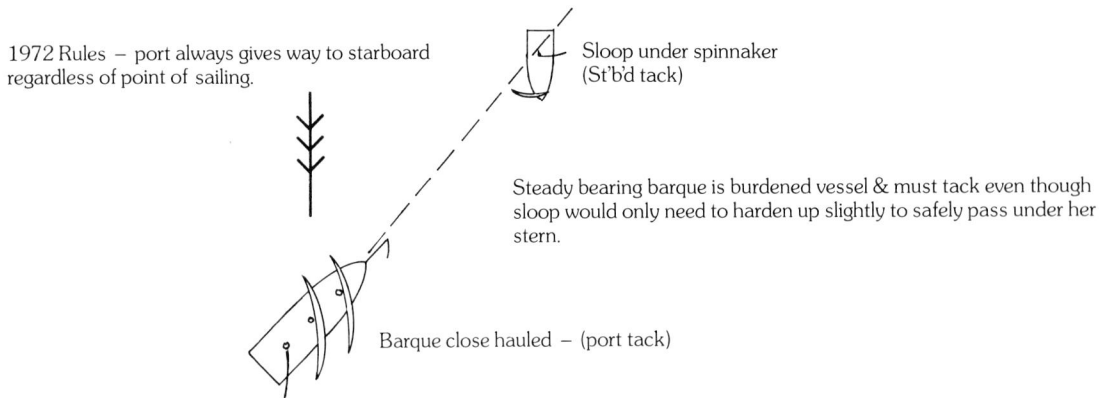

1972 Rules — port always gives way to starboard
regardless of point of sailing.

Sloop under spinnaker
(St'b'd tack)

Steady bearing barque is burdened vessel & must tack even though
sloop would only need to harden up slightly to safely pass under her
stern.

Barque close hauled — (port tack)

Figure 8-4 — Rule of the Road Problems — Close hauled port tack

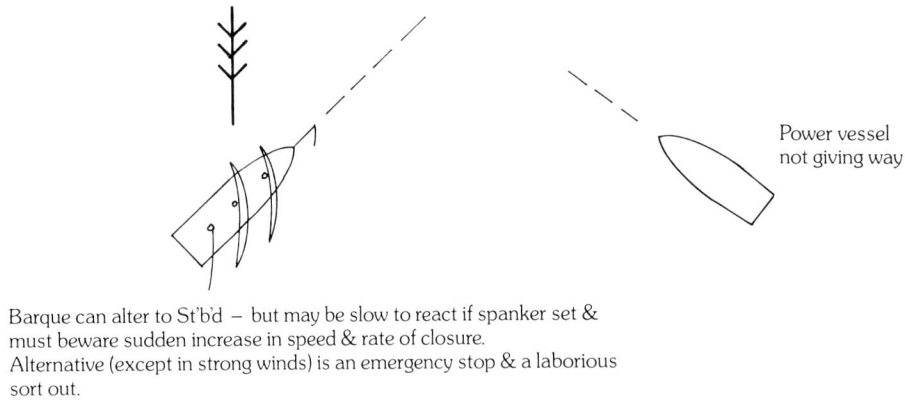

Power vessel
not giving way

Barque can alter to St'b'd — but may be slow to react if spanker set &
must beware sudden increase in speed & rate of closure.
Alternative (except in strong winds) is an emergency stop & a laborious
sort out.

Figure 8-5 — Rule of the Road Problems — Close hauled starboard tack

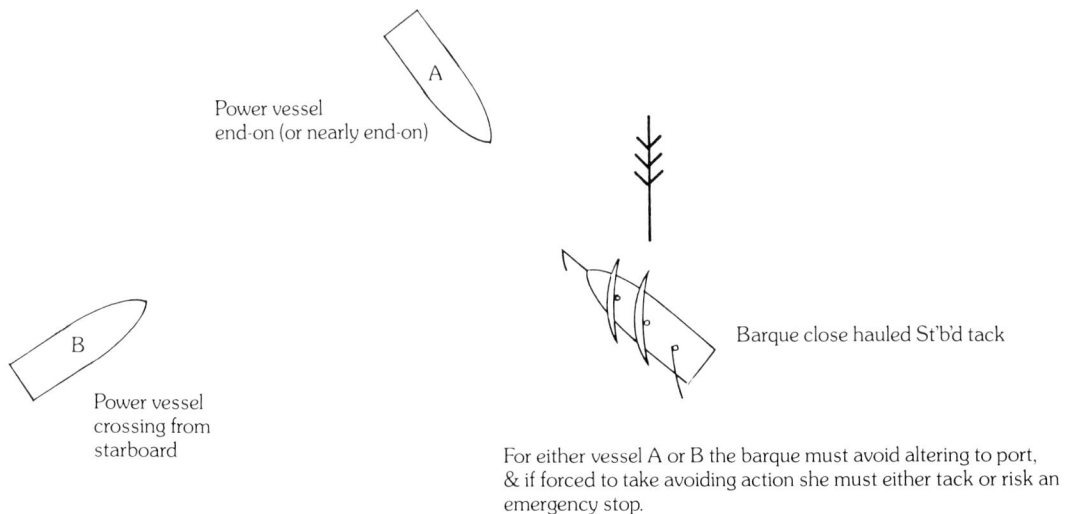

A

Power vessel
end-on (or nearly end-on)

B

Power vessel
crossing from
starboard

Barque close hauled St'b'd tack

For either vessel A or B the barque must avoid altering to port,
& if forced to take avoiding action she must either tack or risk an
emergency stop.

a close pass for photographs. This is particularly the case with small yachts in areas such as the Solent.

Separately the number of occasions under sail when one is embarrassed by a power-driven vessel, either by ignorance, by error, or by sheer bloody mindedness, is increasing rather than decreasing. As Captain Willoughby says in his *Square Rig Seamanship*: '*In all circumstances it is unwise to assume that your vessel has been identified as a sailing vessel with 'stand on' rights . . .* '

There is a well-known doggerel about Michael O'Day '*who died defending his right of way,*' which should serve as a warning that the rules cannot be followed blindly, and without any regard for common sense. By contrast the three scenarios which follow illustrate a more cautious approach, which could be termed defensive driving. As part of this it is always best to set the vessel up to provide maximum manoeuvrability, and minimum lag in response to emergency helm movements. This means that it may be prudent to take in the spanker early (see Chapter 4), to improve your ability to bear away quickly.

Crossing—from starboard

This is a most common event even when displaying the optional red over green, or in broad daylight!

a) **Running free:** Not a real problem, you can easily alter under his stern, or parallel while you prepare to tack.

b) **Close-hauled port tack:** Bear away under his stern, but beware the sudden increase in ship speed and closing rate. Alternatively, if you have not left it too late (and if the wind is less than @20 knots) do an emergency stop by throwing her up into the wind. The risk of this emergency stop is that if the wind is too strong you can overstrain the rig.

c) **Close-hauled starboard tack:** Run off on to a parallel or opening track, and prepare to wear. Alternatively, under **Rule 17a(ii)** make an early tack; however, this is less practical in the larger vessels, as preparing to tack takes longer, and it therefore requires almost uncanny anticipation.

Crossing—from port

This is less common because there is no possibility of confusion as even under **Rule 15** the power-driven vessel would be the give-way vessel. But in many ways this creates worse problems, not least because this is a definite rogue who cannot be relied on to act in a logical or seamanlike manner.

a) **Running free:** As before not a real problem, you can easily alter under his stern, or parallel while you prepare to tack.

b) **Close-hauled port tack:** Run off on to a parallel or opening track, and prepare to wear.

c) **Close-hauled starboard tack:** Bad news! If you bear away to port under his stern there will be a sudden increase in ship speed and closing rate. Further this is the situation that **Rule 17c** warns about (for power-driven vessels), because if your opposite number now chooses to wake up and alters to starboard—then you may be caught beam on in an irretrievable situation. Alternatively if you have not left it too late (and if the wind is less than 20 knots), you can try the emergency stop (as above), with all its associated risks and problems.

Head-on situations

Relative velocity makes things happen very quickly here, giving you much less time to react than in the crossing scenarios.

a) **Running free:** No problem here as you can alter to starboard easily and quickly without the need to trim the yards.

b) **Close-hauled port tack:** Again, you are free to fall off to starboard, at least initially without even the need to trim the yards.

c) **Close-hauled starboard tack:** In a **Rule 14** situation both vessels would alter to starboard, and you should expect the burdened power-driven vessel to act in this sense (though possibly too late, or insufficiently). Thus you should avoid an alteration to port—however attractive it might appear. Since you are already close hauled, a bold alteration to starboard is not an option here without putting in a tack or carrying out an emergency stop. If you decide to tack you would be most imprudent not to have the engine(s) ready on stand-by to guarantee achieving the tack.

Shipping problems—motor-sailing

In many ways this is much more difficult than being under pure sail. The engine(s) only provide a marginal increase in manoeuvrability compared to pure sail, but now the vessel is treated as a power-driven vessel in accordance with **Rule 3c,** and must give way accordingly. What is relatively easy for a small fore-and-aft yacht is a considerable problem for a square rigger. The time needed to manoeuvre, and give way, particularly if a tack is required, can be a real barrier to making progress in a heavy shipping situation, and in these circumstances this factor alone may make motor-sailing an unattractive option.

A further problem is that in daylight it is extremely difficult to find anywhere conspicuous to mount the cone (apex down) as required by **Rule 25e.**

Chapter 9

ROUTINES AND TRAINING

Watchkeeping

THE ARGUMENT over which watchkeeping system to maintain still rages strongly. Watch-and-watch (one in two) is traditional, but feedback (both from cadets and officers) indicates that it puts too much of a premium on sleep, and that today's shorter cruises do not allow people nearly enough time to adjust the body clock.

One specious argument used in favour of watch-and-watch is that it keeps people occupied. Admittedly, it does keep everyone on duty for at least half the day, but that is cosmetic only, because it also introduces the management problem of finding employment for everyone in these large watches.

Those who go to sea professionally in the Navy, or the Merchant Service, become inured to the routine of watchkeeping over a number of years, and for those starting in the business, it is such a part of the fabric of life that it is accepted without question. However, sail training vessels are different, and the command need to put aside their own preconceived ideas, and seriously consider the type, sex, age, stamina, and nautical ambitions of their trainee voyage crew. In the majority of vessels the trainees are on for the experience, and few of them will be tempted to take up what is regarded as a demanding career in a declining industry. The vast proportion will be totally unprepared for constant exposure to fresh air and the elements, and the very young (under 17) in particular can quickly become overtired, and overwhelmed.

Tables 1 and 2 illustrate the problem, using the following assumptions:

- 30 minutes is the standard trick length, because lookout efficiency falls off rapidly after any longer.
- The 'mother' watch (duty messmen, or cook's assistants, etc.) consists of a total of four people from the ship's company, drawn equally from the watches.
- 'Gophers' do below-deck rounds, shakes, and other duties for the OOW.

The issue is arguable, but it is considered that it is better for the trainees to be divided into more (but smaller) watches, with the duties within each watch being shared round more regularly, and therefore more obvious employment. Having only two watches does mean that each watch is on duty more frequently, but there are so many people in the watch that the system is swamped, and there are too few tasks to prevent boredom. Young people today quite sensibly resent doing things for which they cannot see the point. The two-watch system thus tends to be viewed by them as a large loss of sleep for little or no reason, except 'being there.'

Table 1: 28 Cadets—Four Watches of 7—Forenoon Watch Roster:

	Helm	Starboard	Port	After	'Gopher'	Spare	'Mother'
		—— Lookouts ——					
0800-0830	A	B	C	D	E&F	Nil	G
0830-0900	F	A	B	C	D&E	,,	,,
0900-0930	E	F	A	B	C&D	,,	,,
0930-1000	D	E	F	A	B&C	,,	,,
1000-1030	C	D	E	F	A&B	,,	,,
1030-1100	B	C	D	E	F&A	,,	,,
1100-1130	A	B	C	D	E&F	,,	,,
1130-1200	F	A	B	C	D&E	,,	,,

Table 2: 28 Cadets—two Watches of 14—Forenoon Watch Roster:

	Helm	Starboard	Port	After	'Gopher'	Spare	'Mother'
		—— Lookouts ——					
0800-0830	A	B	C	D	E&F	g-1	M&N
0830-0900	g	h	i	j	k&1	A-F	,,
0900-0930	F	A	B	C	D&E	g-1	,,
0930-1000	1	g	h	i	j&k	A-F	,,
1000-1030	E	F	A	B	C&D	g-1	,,
1030-1100	k	1	g	h	i&j	A-F	,,
1100-1130	D	E	F	A	B&C	g-1	,,
1130-1200	j	k	1	g	h&i	A-F	,,

In a four-watch system people have sufficient rest; it is practical to have a full programme of off-watch instruction in the daytime; there is a greater element of competition (more watches); and the Captain has ability to call out the watch below (or even all hands) at dead of night with a clear conscience. The well-known Captain who used to call all hands at 0200 regularly—despite working a two-watch system—must be considered the exception!

If it is decided to work a two-watch system, it is recommended that in summer time the Scandinavian system of watches is used in preference to the traditional British system. In this alternative system the First, Middle and Morning (the night watches) are unchanged, the Forenoon is extended to 1300, and an extra long Afternoon absorbs the Dogs. These long day-watches mean that large watches are more likely to get through their roster in a watch, and also allow a more settled rotating programme than is achieved by the short Dogs. However these long watches cease to be an attractive option once the nights draw in.

Joining routine

The first day is always a rush, and to some extent the trainees are swamped with information. This is inevitable, and regardless of appearances to the contrary some of the information is retained! Ships vary, but the basic essentials are as follows:

a) Division into watches.

b) Introduction to professional crew.

c) Safety harness brief.

d) Introduction to climbing aloft ('up & over').

e) Explanation of basic commands.

f) Simple sail & bracing drill.

g) Explanation of routines worked above & below decks.

h) Above & below deck tours by watches.

In addition, before sailing a further four items **must** be covered:

● Manoverboard procedure.

● Fire drill.

● Abandon ship drill (including use of lifejackets).

● Steering brief.

Many Captains consider this latter safety and steering brief to be so important that they normally do it themselves, and would only delegate it in exceptional circumstances.

Meal times

There are occasions when it appears that the entire ship's programme revolves around meal times! The fact is that feeding a large number of people from a small galley is a major evolution, and shifting meal times at short notice has a knock-on effect on the cook's plan of the day. The cook already has an unenviable and difficult job, so it is important for the command to plan ahead, and where possible avoid any clashes with meals.

Maritime pollution regulations

These regulations are generally known as Marpol 73/78, and the fines for breaches are severe. It is therefore important that prominent notices are put up, and clear briefings given to all on board to ensure that the provisions of these regulations are obeyed. Apart from anything else, training vessels should be setting an example.

Crew stations

In running a modern sail training vessel, sheer economics dictate a large number of trainee crew on board for a short time (typically one-three weeks), and this in turn requires good management and a systems approach, to maximise training value, and minimise confusion.

All such vessels are greatly over-manned by comparison with the last days of commercial sail, but that is not very relevant as historical research is not the object at hand. What is relevant is that it is not practical to use the old system of having a cadre of trained manpower, relying on the others to pick it up over the course a couple of 100-day voyages to Australia (or South America) and back.

Schooners can avoid crew stations, because they have relatively few to fill. But for square riggers this is more difficult, as the requirement for teamwork and co-ordination is part of their attraction, and this tends towards allocated crew stations for most evolutions, at least in the early days of a cruise. Some vessels, in a desperate attempt to be non-military, do not have stations, but the end result is that the unfortunate trainee crew are herded about like cattle, with no time to learn any particular task, or to look around and take a wider interest. The ability of such vessels to react in a real emergency must be questionable in the extreme.

A large crew which is correctly organised can indulge in all the old style naval evolutions such as short tacking inshore, simultaneous setting/furling sails, and manoverboard drills. All of these totally involve the entire crew, thus both providing real training, and helping build crew morale. The sense of achievement for the trainees is very positive, as they realise that they are changing from hamfisted bunglers into a slick well-oiled machine.

Training programme for cadets

For civilian vessels the training programme on board should be written around the RYA Offshore scheme (or similar for other nations), unlike the government-funded Naval or Merchant Service vessels, which will carry out longer syllabus professional training. There is an element of lecturing in all such programmes, but it is not easy to put over any subject in a less than ideal environment—such as a messdeck—particularly when the instructional aids are also somewhat primitive. The only vessel whose design really addresses this side of training is *Spirit of New Zealand*. She has her dining hall cleverly laid out in the stern with tiered seats, and all the instructional aids are built into the deckhead, enabling it to double as a very adequate lecture theatre. It must be said that motion at sea is greater than the more traditional near midships position, thus reducing its 'popularity' as a dining facility! However, such criticisms should not obscure the fact that this is a very interesting and innovative idea.

In only a week—starting from scratch—a trainee can become quite useful, and in a fortnight those with any aptitude can achieve RYA competent crew standard. Intelligence and/or some sailing experience (dinghies) are helpful, but not essential, as motivation is far more critical. The message to sell is that once they do learn, the permanent crew will shout at them less, and everyone can relax and enjoy life. However, it must also be borne in mind that to a great extent those that benefit the most are not the 'stars,' but those that really struggle, and make little apparent progress, and the training programme must reflect this factor. It is all too easy to fall into the trap of believing that sail training is about producing a new generation of seaman—for which there is only a tiny requirement—whereas in fact it is merely a tool to achieve character development in the widest sense—for which there is a very large market/requirement.

Unlike normal Outward Bound, the sea by its very nature provides quite enough tests of character, and there is no need to invent rivers to cross and such like. Although most will be on their one and only sea voyage, there should be no alteration in the uncompromising professional standards demanded, as part of the training value is learning to live in an environment where close enough is **not** good enough.

All the permanent crew should be to some degree directly involved in providing instruction to the cadets.

Training programme for afterguard

Most training for the regular crew is 'on the job' training, and this is all right as far as it goes. However, it will generally be found that things work better if the regular crew (and any voyage officers) are given some opportunity to carry out basic manoeuvres under sail. They will enjoy it, and having looked at the problem from a different angle, they will be better placed to carry out their normal duties. In addition in due course some at least will move on and up the ladder, and the earlier that they start to gain ship-handling experience, the better. For one young Captain, the result of serving under old school Masters (who believed that only they should handle the ship), was that his first proper manoeuvre under sail was to get under way from a kedge (at night), with a crew of whom half only spoke French.

Those vessels which do long cruises—or have cadets working up over a number of cruises—should also involve them in basic shiphandling. Tacking the vessel is a great kick for the trainee concerned (particularly the quieter ones), and in addition those with real aptitude raise the level of training above the mundane, which is good for the ship's officers.

From a safety point of view the ideal is that everyone in the hierarchy should have someone who could replace him in an emergency.

Women at sea

One of the advantages of using square riggers for sail training is that the work entailed in running such vessels emphasises skill and teamwork, rather than individual brute strength. This means that the women can compete with the men on equal terms.

Some say that a mixed crew humanises the boys, whilst toughening the girls, and perhaps there is some truth in that. Certainly people seem to drive a mixed crew harder than an all-female one, and seem to be less tempted to take in sail early. It is absolutely essential to have an afterguard that contains both men and women. Men do not get the best out of difficult girls, and cannot cope with tears, whilst bolshy teenage boys do not respond well to women in authority.

The permanent crew should not take advantage of their position to have affairs with trainees (regardless of age) because it interferes with the training programme. This should be no hardship, as for the (normally male) crew member it tends to be an act of rather callous/amoral self-indulgence. As regards relationships within the professional crew, the Master has the same problem as in any other ship—namely ensuring that they do not undermine the chain of command, and that a reasonable coherent team is maintained.

Foreign trainees

Given the demographic trough, and the ease of travel, it is likely that the numbers of European trainees will increase in British vessels. In addition, in the Tall Ships Races there is always a crew exchange. Although many of the cadets will speak reasonable English, this is unlikely to include a maritime vocabulary. It can be helpful if the deck officers learn a few basic sea terms for explanations. Selecting one of the Scandinavian languages, or German, means that you can master terms common to the North European languages. A short list of German and Norwegian equivalents is provided at Appendix 3, and it shows that the terminology is remarkably standardised. However beware of the French, because their system is very different, with for example the 'Misaine' being the foremast, instead of the mizzen as one might expect!

Royalist—*stowing the mainsail—a great advantage of a square rigger is that there is always work to do, even if it is only practising furling in a flat calm. Note the reefing points hanging down on the fore side of the topsail (Author).*

Chapter 10

SAFETY ON BOARD

Safety philosophy

Running a sail training vessel requires the command to maintain an environment that is both safe and challenging. Whilst nothing at sea can be entirely safe, it is believed that these two requirements are in no way mutually exclusive, and that a good measure of success for a long-term sail training vessel is to examine its safety record. A symbiotic relationship between apparently conflicting factors is by no means unknown, indeed in the Navy it is generally reckoned that 'a happy ship is an efficient ship' (and vice versa). If a good standard of safety is to be achieved, then some measure of discipline is vital, and it is in fact considered that this discipline is an essential element of sail training. The degree and formality of discipline is open to wide interpretation, but it cannot be too highly stressed that notwithstanding what may (or may not) be true ashore, at sea a policy of unrestrained 'free will' is naught but a recipe for disaster.

It should also be clearly understood that major accidents do not have single causes, they are the result of a chain of minor cumulative and inter-connected factors. Every safety course stresses that all that is necessary to prevent an accident is for one person to break that chain. As far as the professional crew are concerned, it must be drummed into them that regardless of their strict job description they must look around constantly, and **'think ship safety.'** For example, the cook should be able to cast an intelligent eye around on deck, and likewise the bosun should be able to recognise a potential accident in the galley.

SAFETY IS NO ACCIDENT

Peer pressure

Although peer pressure can be a useful tool in encouraging cadets to do things that are frightening, unpleasant, or tiring (or a combination of the three), it does have a down side in that it is possible to push people beyond their capabilities. This is a problem area that the older and more experienced of your afterguard will take care to avoid, both by sensible supervision and by setting a good example. Unfortunately it is unlikely that all your crew will be of this calibre, and there will be at least one under 25 year old as a bosun's mate or such like.

These young people are very enthusiastic and keen, but (particularly if they are men) they are prone to revelling in their perceived superior status and knowledge—compared to the trainees. This is in itself fairly harmless, but if unchecked it can lead to dangerous showing off such as sliding down backstays, and then this peer pressure can lead to the less capable cadets attempting to emulate this derring-do with possibly fatal results. Being a crew member of a sail training vessel is a responsible job, and anyone who continues to allow his ego to outweigh his sense of responsibility must be allowed to depart to a less 'demanding' job.

Safety equipment

All on board should be issued with a lifeharness as an item of personal safety equipment. These must be carefully checked and serviced before each issue, and can only be repaired by the manufacturer (for certification). Care for these items must be emphasised as being in their own interest.

No other items are likely to be on permanent issue, but the same degree of respect should be inculcated for all lifesaving/ship safety equipment. Misuse, or maltreatment, of any of this gear cannot be tolerated, and any such stupidity should be treated as an offence likely to endanger life.

Sorlandet — *stowing main topgallant* — Norwegian cadets showing how it is done, and doing it quickly. Note that once in position they have correctly clipped onto the after (safety) jackstay, NOT the backwire, and that they are relying for security on their arms, and a good body position (Author).

Health and safety at work

A considerable body of legislation has been enacted to lay down minimum standards of safety and protection when engaged in normal ships business or in maintenance activities. This covers matters such as use of ear defenders, goggles, gloves, etc., and the enforcement of these safety standards should be second nature to a professional seaman. However, trainees and young seasonal volunteer workers are unlikely to have this degree of basic knowledge and will need careful supervision and instruction. These days you cannot just arm someone with a chipping hammer and leave him to get on with it; not unless you wish to risk a heavy lawsuit for negligence.

In addition to this legislatory detail there a number of measures that are recommended to improve safety, and reduce the chances of injury:

- Rings and earrings can both get caught in ropes or rigging, and the resultant injury will at best only be bloody. Therefore neither of these should be allowed to be worn on board under any circumstances. Covering over with tape is insufficient, and those rings that cannot be removed normally should be cut off, as it is better to damage a ring than to lose a finger.
- Necklaces, bracelets, and watches can also get caught, particularly aloft. Only permanent crew members have any need to use a watch, and no-one needs the other items of decoration, so they can be left off for the duration.
- Notwithstanding the macho image factor, bare feet are an invitation to injury. Non-slip shoes (or boots) should be mandatory at all times on deck, and aloft.
- Really long hair can get caught in a block, with the risk of scalping, and in such case it needs to be adequately secured.
- Obviously nothing should be taken aloft that is not on a lanyard. People tend to notice knives and tools, but time after time the pocket camera is forgotten, only for it to come crashing down from aloft. Fortunately they normally smash on deck, but even the smallest camera can be an effective missile after falling 50 ft or more.
- Gloves should not be worn aloft, or when handling ropes, as they can get caught and cause an accident. Obviously gloves may be worn at other times to keep out the cold. They may also be used to protect the hands when splicing and working with wire.

Climbing aloft in harbour

The normal policy in harbour should be to climb up the outboard side where possible, on the grounds than any fall will end up in the water, rather than on the concrete or stone jetty. It will not totally prevent injury, but it could make the difference between a serious accident and a fatal one.

Back wires

Some vessels have back wires (or ropes) rigged between the lifts and the chain tie on their yards. Those going out on the yard duck under the wire, and can then clip on to it while they move out into position. This is slightly easier than clipping on to a jackstay, and the wire at their back provides added confidence for trainees. The argument against this is that anyone falling will fall further than if attached to the solid jackstay, and the shock load on the wire may pull others off the yard. If such a wire is fitted then the potential shock loading must be carefully calculated and recorded, the exact layout fully documented, and an example professionally proof tested. The reason for all this is that this is a safety feature, and therefore it is essential to pre-empt any charge of negligence in the event of a failure, and resultant accident. Lack of documentation might well invalidate normal ship's accident insurance in such a case, and since litigation is now the norm, that could place the Master in a very difficult position.

Futtock safety wires

Most people feel vulnerable going over the futtock shrouds for the first time, particularly while they are not clipped on and have to lean outwards to reach over the top. For about 5 per cent this feeling is so severe that they freeze up, or panic, and extricating the more acute cases can be difficult. To improve both confidence, and safety, some vessels have incorporated simple 'futtock safety wires' into their design. These wires are separate from the standing rigging, and shackled to the bottom of the futtock shrouds, and at a point one metre up one of the topmast/topgallant mast shrouds. The trainee can thus remain clipped on while climbing up, and can clip on before descending. In any fall it would be normal for the rate of descent to be markedly reduced by the person concerned (and the supervisor), so that the only damage was bruised pride! Exceptionally a cadet who totally blacked-out would deadfall enough distance to (probably) suffer kidney damage; however, countering that problem presents a technological challenge requiring the development of new equipment, such as shock-absorbing harnesses.

Ratlines

The lashings on traditional rope ratlines need a lot of maintenance, and it is a good idea to replace those on the lower shrouds with ratling bars. These ratling bars should consist of a length of wood, backed with galvanised (or stainless) angle iron, and secured to the shrouds either with wire seizings, or bulldog grips with the nut heads recessed into the wooden bar. The result is immensely strong and long lasting, which is important as it is the lower ratlines which absorb the greatest wear and tear.

This system cannot be continued on the upper rigging, because the yards bear on the rigging. But where they are used a further advantage of these ratling bars is that they inspire confidence in the trainees, and initially encourage more of them to go aloft than would otherwise be the case. As always, if you can get them aloft once, then the battle is almost won. Checking the seizings on the rope ratlines remains a regular weekly chore for the bosun. Boring, but essential.

Hypothermia

In our centrally heated urban society exposure to the elements can be very limited indeed, and most trainees are totally unprepared for the lack of warmth on even a mild night in spring, or autumn. Those coming on watch need to be physically checked for the first few days, to ensure that they are wearing sufficient warm clothing to last their four hours on watch. This is made worse by the fact that many will only have nylon socks and underclothing, and these provide almost no insulation against the cold. For this reason, and because of the fire hazard (nylon welds to the skin), joining instructions should therefore strongly recommend woollen (or cotton) clothing for wearing next to the skin. Seventy per cent of heat loss is from the head, so a supply of woollen hats available (on sale) on board is a very sensible idea, which may also raise some money for the vessel. Few of the trainees are likely to have brought one with them even in the coldest of weather.

It must also be stressed constantly that anyone who gets wet must change into dry clothing, once the physical work is done. Keeping people standing around in wet clothes is bad practice, and is not to be regarded as part of some toughening process. Some trainees will manage to run out of dry clothing, consequently it is prudent to have a couple of sets of 'survivors' kit' available. These kits should consist of a large thick sweater, and heavy working shirt and trousers, and they should be selected for warmth rather than style.

Even with all these precautions there is still the risk that some cases of hypothermia will slip through, but provided these are caught early the victim should suffer no ill effects. Permanent crew members must be fully familiar with the symptoms and the immediate treatment, and voyage OOW/watch leaders should also be carefully briefed. The basis of treatment for simple cases is dry clothes and a sleeping bag, and then a long wait with a nominated person to 'observe and monitor,' until the patient recovers. It must always be remembered that constant vigilance is essential in cold weather, because **cold can kill.**

Watertight integrity and damage control

Sail training vessels do not enjoy the same degree of watertight sub-division as warships, but adherence to sensible rules can enhance their safety factor, and these rules should be laid down in ship's standing orders:

- Doors should be marked up and carefully maintained.
- There should no longer be any opening scuttles in the ship's side, but all the deadlights need checking regularly, as they are irresistible to 'fiddlers.'
- The permanent crew should all be trained in the use of shores and wedges (Naval vessels only).
- A plan should be exercised to use bunks etc to reduce free-surface effect in the event of major flooding. Remember if the area is cut in half, then the effect is reduced by 75 per cent.
- All deadlights and watertight doors should be shut in conditions of reduced visibility.
- Heavy weather preparations should include shutting down vents, and closing all but essential upper deck doors and hatches.

Fire-fighting

Fire-fighting is the responsibility of the professional crew, but only a few key people will have the skill or training to attack a serious fire. The prime responsibility of the crew is moreover the safety of the cadets, and the saving of the ship must come second. A key element of Captain's rounds is checking for fire hazards, and ensuring that all equipment is in date. Portable fire-fighting equipment should also be checked for correct pressure/weight, in case of inadvertent discharge (particularly after a bout of heavy weather).

Larger vessels will have full suits and breathing apparatus (BA), and those trained on BA must be regularly exercised in the techniques for finding and recovering someone in a smoke logged compartment. Covering over the BA facepiece provides adequate simulation for this drill.

Fire-fighting is a skill that requires good basic practical training, and (ideally) regular refreshers. It also requires the command to have fire control plans worked out, and prominently exhibited. These plans should cover the problem of excessive water creating the extra hazard of **free surface effect.**

One Master visiting another sizeable sail training vessel was horrified to witness a BA rescue exercise carried out by two scruffy young so-called 'professional' crew-members, neither of whom had ever done any sort of fire fighting course, or was at all aware of even the most elementary techniques. For a start both went in wearing thin soled sailing shoes, and nylon underclothing! They could not understand that their mistakes, and lack of knowledge, meant that their sole contribution to a real incident could only have been to add themselves to the casualty list.

Radio and radar transmission hazards (Radhaz)

There is considerable and justifiable fear about the long-term effects of exposure to high-powered electro-magnetic radiation of any kind. However, the relatively low power output of commercial 9 GHz band radars means that the greatest danger from the aerial is being hit by the rotating head—if working too close. In fact the inside of the display is much more lethal, as there is some very high-power electricity inside, and amateur maintenance is definitely not advised. Fortunately, digital electronics normally need spare circuit boards, rather than screwdriver adjustment.

MF radio aerials route to the backstays, and no transmissions should be made with men aloft. A warning notice should be permanently placed near the set, and a further one in the rigging when transmissions are being made.

The galley

The galley is a dangerous place for the unwary, even in harbour. In normal working hours the cook will be around to supervise, but in the silent hours there is a need to make hot drinks. Therefore a routine needs to be developed to allow this requirement to be met without boiling water flying everywhere! Some smaller vessels require cadets to put on a foul weather jacket before going into the galley to make drinks, as the jacket will at least limit any scalding.

The engineroom

The engineroom is not a place for unsupervised trainees. It should be firmly placed out of bounds, except for instruction, or cleaning routines, as directed by the Chief Engineer (or one of his men).

Anchors and cables

A power-operated anchor windlass is potentially dangerous, and it should never be operated except by trained personnel. This area is normally the province of the boatswain, and when anchoring he must ensure that the trainees are fully aware of the dangers both up on the cable deck, and down in the cable locker.

Boat routines

Anchoring off (or being at a mooring) means running liberty boats, and drink and boats mixed have caused more accidents over the years than almost anything else. Better to leave a bolshy drunk ashore (possibly in friendly police custody), than to put other lives at risk in a boat trip. Everyone in a boat should wear a lifejacket at all times, regardless of the conditions. There can be **no** exceptions to this rule.

Hands to bathe

There was a case some years ago of a vessel having 'hands to bathe.' After a while one trainee dived from the lower rigging, and eventually things escalated to diving from an upper yard, with the end result that one youngster misjudged it and was killed. A classic case of peer pressure going wrong, it also demonstrates what happens when supervisors shirk unpopular decisions—being unwilling to 'spoil the fun.'

Some years later in another vessel the Captain emerged on deck to find a similar situation developing, and only after he had ripped the OOW to pieces, did he realise that this previous disaster had already faded from the collective memory. In their own way these are both cautionary tales, because together they show not only how easily an accident can occur, but also how easily it can be repeated.

'Hands to bathe' is popular and good fun for all, provided that it is properly supervised by an officer on deck, and that there is a safety boat in the water (under oars/paddles) manned by at least two. In the tropics it is also essential, to have a shark lookout up in the rigging, and to brief the swimmers not to get too far from the ship. Sharks seem to be attracted by people splashing in the water.

First aid

Prevention is better than cure, so this item is placed last; however, if there is an accident it can be of critical importance. There is no substitute for professional first aid training, and the command should take careful note of all qualifications, not just for the permanent crew, but also amongst the trainees. There is little point the Master and Mate blundering around with the aid of the *Ship Captain's Medical Guide*, when there is a fully-qualified doctor (or nurse) available.

That said, first aid can work wonders, and in serious cases never mind the what it looks like, because the cosmetic side can be left to the surgeons ashore—provided that you give them all the pieces.

Royalist — *ratling bar detail — the galvanised angle iron backing, and the bulldog grip attachment to the shroud are shown clearly (Author).*

Chapter 11

MANOVERBOARD UNDER SAIL AND UNDER POWER

MANOVERBOARD (MOB) is a constant fear in any vessel, and although the actual incidence is low, reactions have to be regularly exercised as there is no margin for error. The most risky work in a sailing vessel is not in fact working aloft, but being out on the bowsprit in heavy weather. This is especially true in a sail training vessel, because the cadets have little appreciation of the power of the sea when you ship it green, and they are unprepared for the violent vertical acceleration forces generated by the motion. This motion tends to be at its most extreme when using the engines to power through a head sea.

A further problem is that generally you do not have a good view of work on the bowsprit from the conning position. In order to ensure that risks are minimised, and that any MOB is reported immediately to the command, you must ensure that all work out on the bowsprit is closely supervised. Initial MOB reactions—detail lookout, take a fix (Decca and GPS have MOB fix buttons), call the Captain, etc.—are unchanged in sail or power, and it is not intended to reiterate them here, but to concentrate on techniques and aspects particular to sail training vessels.

Manoverboard liferafts

Survival time in the water in the Northern European and North Atlantic operating area is not that long in summer, and in a hard winter it reduces dramatically. One good idea which originated with the barque *Gorch Fock*, is to have a liferaft (with an automatic light) located on either side of the quarterdeck, and specially fitted with an extra large weighted sea anchor to reduce drift.

In the event of a MOB, the officer of the watch immediately launches one of these liferafts, cutting the painter when it inflates, and the person in the water knows that life depends on reaching that liferaft, because once there he could survive for several days in a liferaft (if necessary), and a liferaft is far far easier to spot than a man's head (or a woman's for that matter), all of which gives a tremendous boost to critical life-sustaining morale. If the MOB is seen to reach and enter the liferaft, then the time element is less critical, the pressure on the ship is reduced, and thus allowing a deliberate manoeuvre, and reducing the chance of error.

Two liferafts should be allocated to this role, since there is always the possibility that one may not work. These MOB liferafts are part of the normal complement, but since that complement includes a massive excess in capacity the use of a liferaft in such an incident seems entirely seamanlike, and indeed the idea has been taken up by several other vessels. The DTp lifebuoys are all very well, but like the infamous World War II Carley rafts they only help to keep the person afloat, while completely failing to provide for the first priority for survival—protection—and contributing very little to the second priority—location.

MOB markers

Any extra markers that can be dropped to assist relocation of the MOB are invaluable. Standard smoke floats are good, as is a danbuoy fitted with both a light and a radar reflector. Some ocean racing yachts have a very slick arrangement, and carry their danbuoy in a chute at the stern, with a launch mechanism that can be activated by the helmsman.

Manoeuvres under sail

When manoverboard is exercised it is often also used as a sailing evolution utilising the old 'reach—tack—reach' manoeuvre followed by heaving-to. Whilst this is all good crew training, it is not a viable tactic in reality as it is too complex and too time consuming. In the water would be a frightened, untrained, youngster, with very limited self-preservation ability or confidence. Intricate sailing manoeuvres can go wrong, and

Lord Nelson — *manoverboard liferaft installation — this is a good example, with the traditional lifebuoy also close at hand (MAX).*

they also involve significantly opening the range from the MOB. This is bad enough with one of the smaller vessels, whose average speed of 6 knots equates to 200yd/180m per minute, but once you get up to the larger vessels running possibly at 12 knots (400yd/360m per minute), then your chances of remaining in touch are very small.

The greater the range, the greater the difficulty of maintaining visual contact, and also the morale of the person in the water is not improved by the sight of his vessel continuing to sail away. Add to that the problem of closing everyone up (particularly at night) and the increased chance of error (owing to stress), and that evolution drill looks less and less attractive. The basic options for manoeuvre given here, have by contrast been deliberately kept as simple as possible.

Square rigger (fair weather)

In less than about a 20-knot wind speed, execute an **emergency stop**—regardless of the point of sailing. To achieve this throw the vessel straight up into the wind to stop her, haul the main yards, and then sort yourself out. Hauling the main yards ensures that whether or not you got through the wind you will end up hove-to. With a responsive vessel and quick reactions, you could end up very close indeed to the manoverboard. At the very least you will not be opening, and will have a good datum position, which is vital at night, because visual contact will have been very quickly lost.
- **In visual contact:** Remain hove-to and depending on distance recover by seaboat, or heaving line.
- **Lost contact:** This will certainly be the case at night, but may also occur by day. Whilst still hove-to you need in short order to furl all (square) sails, carefully check that there are no lines in the water, then start main engine(s), and man and lower the seaboat. Once underway you should keep the seaboat close on your quarter (ready to pounce if MOB sighted), while you conduct a square search starting on the datum. Early assistance by other vessels or a helicopter will greatly increase the chances of success. Searching from a low platform with largely untrained lookouts is less than ideal, so station your better cadets in the best positions.

Square rigger (heavy weather)

The emergency stop being out of the question, you will undoubtedly lose contact with the MOB. Launching the seaboat only puts other people at risk, so very accurate manoeuvring for a direct ship pick up will be required. In theory you could conduct the search purely under power, but motor-sailing to a 'modified' square search may work better. Conducting an accurate search will be very difficult, and it must be stressed that unless the MOB reaches the liferaft (assuming it deploys correctly) chances of recovery or survival are small by day, and worse by night.

Fore-and-after (all weather)

Even in small yachts, the RYA no longer exclusively favour the 'reach—tack—reach' manoeuvres, and look towards stopping the vessel quickly within striking distance of the MOB using the 'slam stop' (tacking with the jib sheet left fast). In large schooners a similar policy can be used, and the recommendation there is that the OOW should immediately luff-up to kill the speed, and then fall off to 30° off the wind. At this attitude the vessel should be able to maintain steerage way, and be fairly steady, while sail is reduced/handed. This is an easier and less risky manoeuvre than in a square rigger, and would appear to be the favoured manoeuvre in virtually any weather.
- **In visual contact:** The reactions here are the similar to those for a square-rigger, remain semi-stalled with steerage way only, and depending on distance recover by seaboat, or heaving line. You may well not have completed handing sail by the time the MOB is recovered.
- **Lost contact:** This will again certainly be the case at night, but may also occur by day. The difference lies in the actions and manoeuvres after that initial emergency stop. In a schooner you complete handing excess sail in the semi-stalled condition, and then get under way manoeuvring for a final approach at 50° to the wind. Main engines should at least be on stand-by to assist precise positioning, so it is vital to take the time to carefully check that there are no lines over the side. If the weather is suitable for the seaboat, it should be manned and lowered, then held on the quarter ready

to pounce. A practical alternative to handing the gaff sails is to scandalise them by lowering the peak just below the horizontal, thus killing their drive (see Chapter 4). However it has been found that large Bermudan sails (such as the mizzen in the STA schooners) must be handed in good time, as must sails such as jib topsails.

Manoeuvres under power

The reaction here should always be a textbook Williamson turn, limited by the fact that the vessel is comparatively underpowered, and less manoeuvrable in any wind, than a similar-sized power-driven vessel. A Williamson turn does take marginally longer than some of the other potential manoeuvres, but unlike them it sets you up heading down the reciprocal of your track towards the last known position, and is the nearest possible to a 'failsafe' manoeuvre.

The initial turn off heading is normally 70°, before helm is reversed in order to roll out heading back down the reciprocal track. The initial angle off can be quite easily established in a flat calm, but any wind will change the shape of the turn, and this needs to be taken into consideration, and the final roll out adjusted in order to achieve the correct line-up.

Everyone is mentally attuned to 'sail' training; consequently, if you inject an MOB exercise as an evolution while under power on a flat calm day it seems to completely flummox the average young OOW. Since most sail training ships actually spend a significant proportion of their time under power it is important that this power MOB drill is not neglected.

In answer to the inevitable question from the cadets, the turn is named after Captain John Williamson, USNR, who devised it in 1942 while he was a young Lieutenant (jg) instructor in a patrol craft seamanship school.

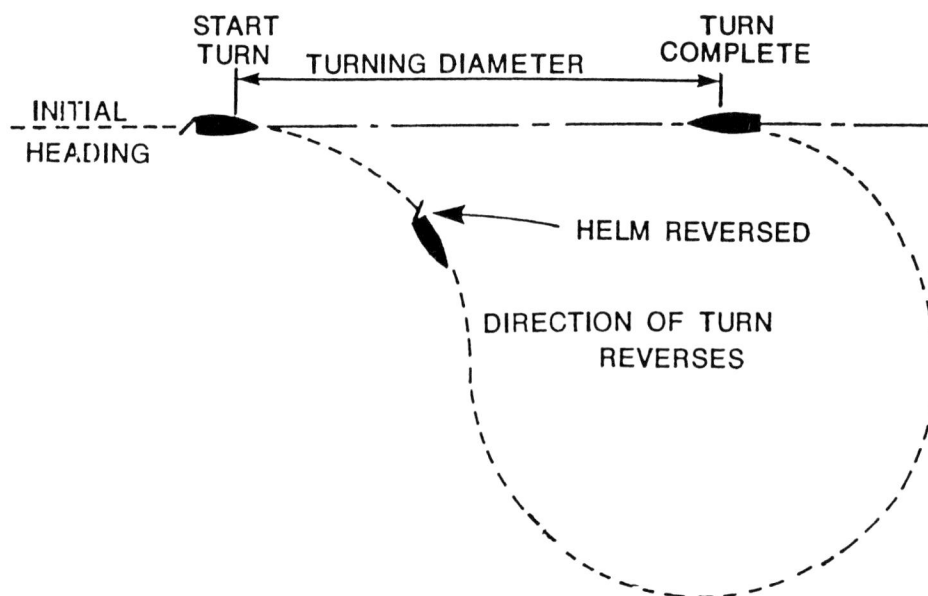

START TURN — TURNING DIAMETER — TURN COMPLETE

INITIAL HEADING

HELM REVERSED

DIRECTION OF TURN REVERSES

Fig 11-1—The Williamson turn

Controlling the seaboat

The crew for the seaboat needs to be at least two strong, fit, and intelligent crewmen wearing proper protective clothing, and fully inflated lifejackets. The decision to risk them in marginal conditions rests firmly with the Captain, and depends entirely on the conditions at the time.

A further argument for using a MOB liferaft is that it makes life easier when recovering the survivor by seaboat. The height of eye from the standard inflatable/rigid inflatable boat is extremely low, and while a liferaft would be visible to the crew, finding a head in the water requires conning directions from the mother vessel. In a

Naval vessel the coxswain might have the luxury of a radio headset, but the rest of us have to fall back on hand-held VHF sets, and these handsets are less than easy to interpret over the noise of an outboard engine. In any case Murphy's law will inevitably ensure that these radios do not work, forcing you to use hand/flag/torch signals, which are only practical with a worked-up team.

Direct pick-up

If the seaboat is used, a transfer from that back on board is relatively easy, indeed you can even hoist the seaboat with the survivor on board. However, if you have manoeuvred for a direct pick-up, you have two problems:

a) You need to position the vessel with the MOB very close on the lee side abreast the boarding ladder/net, and this must be achieved whilst avoiding the risk of running over the MOB.

b) If it is cold (or after a protracted search) it is probable that the survivor will be suffering from hypothermia. In that case not only is time critical for treatment, but the survivor is likely to be at best semi-conscious, and extracting him/her from the water will be most difficult, without seriously hazarding someone else.

In all cases a scrambling net (or boarding ladder extending 2 ft below the waterline) must be rigged, and a tough crewman made ready equipped with a fully-inflated lifejacket and lifeline (and ideally a wet or dry suit). Cold saps the strength in an insidious manner, and even an apparently active survivor can be effectively incapacitated by being unable to grip onto the ladder. A crewman must always be positioned to assist, and to give immediate mouth-to-mouth resuscitation if the MOB is unconscious.

The Mate should be stationed by the pick-up point to provide the Captain with guidance for exact positioning, and to control directly the operation. The recovery crewman once over the side must always remain made fast to the ladder/net, and must be carefully monitored in cold weather to ensure that you do not end up with yet another hypothermia victim. A sodden incapacitated MOB is remarkably heavy and awkward to handle, so unless your recovery crewman is built like a rugby second row forward, you may need to use two people.

You should also note that analysis of deaths from hypothermia has led to the recommendation that in such cases the normal vertical lift from the water may be a 'killer,' and that where possible a horizontal lift should be used. Helo crews use one strop under the arms, and another strop under the knees, but this is not easy to achieve when the rescuer is being battered around by the waves, and the unconscious person keeps slipping out of the strops.

Medical treatment

The main problem is hypothermia and the guidance in the Captain's medical guide is clear, but the actions do need to be exercised before the event, because the treatment must be immediate, and in severe cases there is no margin for error.

Outside assistance

In heavy weather, or at night, the odds are not good, and it is vital that you alert the nearest Coastguard station as soon as possible, and request assistance in your search. You may be lucky and find the MOB in the interim, but no-one will complain when you then cancel your call. However, if you delay until you are certain that you alone cannot succeed, then you have sacrificed valuable time. Time is critical for MOB location and survival, and to improve the chances of success, then other units are needed on the scene, and quickly.

If a warship, or merchant vessel, comes to your assistance, then you would probably be advised to hand the search co-ordination role over to them, because they have more trained manpower, and much better facilities, than a sail training vessel. Equally helicopter and lifeboat crews are specialists, and thus only need the sort of information detailed in Chapter 12.

Annex 11A

ON-BOARD MOB LOCATION DEVELOPMENTS

PROMPTED INITIALLY by the activities of *The Times* newspaper, an extensive investigation of the whole problem of MOB location was initiated in 1989. The basic aim was to improve the position of the Whitbread Round-the-World Race competitors, because they faced a particular problem in the extremes of the Southern Ocean, where survival time is minimal, and outside assistance effectively unavailable.

Trials

After early tests sponsored by *The Times,* an investigative committee was set up, chaired by the race director of the Royal Ocean Racing Club. The results of these practical tests highlighted the fact that—far from there being any cause for complacency—it was extremely difficult to locate a body in the water with existing systems, even with an abundance of highly-trained crew as lookouts, in broad daylight, and in near-perfect conditions.

Furthermore a number of expensive and complex systems proved to be seriously deficient, and/or insufficiently robust, and the danbuoy (including an interesting inflatable version) only lived up to its reputation within about 1,000m.

Electronic location

Moving on from this, the committee developed and tested a short-range VHF (121.5 MHz) personal locator beacon (PLB), and a simple on-board tracking system to enable the crew to locate a MOB quickly, in all weathers—hopefully before the onset of hypothermia. Tests in the rough water of the Portland tidal race showed it to be capable of tracking and homing from up to 4 nm—an enormous advance.

Whitbread race experience

Prototypes were installed in several yachts for the 1989/90 race, and, although there were some problems, the system was used on two occasions, leading to the recovery of a total of three crewmen (regrettably one dead), in difficult circumstances, when location would almost certainly otherwise have been impossible. The old 'Reach—tack—reach' MOB manoeuvre was scathingly rejected by one skipper, as being totally impractical in heavy weather.

By contrast the 'slam stop' was endorsed by its successful execution by one yacht, even though she was surfing at very high speed with the spinnaker set, in strong winds and big seas. She recovered her MOB in six minutes, without recourse to electronic location aids, and was fully racing again within the half hour—thus proving that hi-tech systems are all very well, but there is no substitute for good seamanship!

Future development

Although this is all rather exotic at present, in the longer term it might become practical to issue those working in exposed positions—such as the bowsprit in heavy weather—with a lightweight hazardous duty lifejacket (self-inflating), incorporating a miniature automatic PLB. Relatively short range and life would probably be dictated by the need to contain the false alarm rate, and to frustrate malicious hoaxers. It is appalling, but these hoaxers must be considered—there are all too many of these perverts, whose 'kick' is to activate and divert the rescue services to false alarms, without any consideration for the fact that such diversions put at risk the lives of those involved in real emergencies.

NIGHT SEARCH AND LIFEBOAT CAPABILITIES

Night search problems

NORMALLY at night a man in the water without a light on his lifejacket is most unlikely to be detected, and no-one with any experience of search and rescue can understand why such lights were not made mandatory long ago. For a pure visual search the much vaunted reflector strips are no substitute for a proper fixed light (or a mini-strobe), even when they are in pristine condition. Extensive examination and trials of merchant ship safety gear in Australia has also shown that it is all too common for even this level of strip performance to be dramatically reduced by age, wear, grime, or salt encrustation.

Lifeboat capability

A lifeboat brings to the scene a highly trained, and experienced crew, but until recently their night search capability has been limited to radar, and the 'mark one eyeball,' and they have lacked electro-optional systems—such as forward looking infra-red (FLIR)—that are increasingly being fitted to helicopters (see Chapter 12).

Electro-optics

The military have put a lot of money into developing a variety of electro-optical systems for night operations. These are slowly becoming available for civilian applications; consequently rescue agencies around the world are now investigating suitable systems for lifeboats. In Britain the RNLI has concentrated on image intensifiers combined with illuminator lights, and have achieved some excellent trial results. Using a searchlight and the human eye detection of a body in the water varies from 100 metres at best, down to about 50 metres; whereas with an image intensifier system the results are consistently better—and research and development is still continuing.

One installation under consideration has the camera images remoted to a TV display in the wheelhouse, and this has the dual advantage of allowing all the crew to join the search, and of avoiding the problems that are associated with handheld sets or 'goggles' (poor field of view, eyestrain, and loss of night vision).

Image intensifier based systems, compared to infra-red (or thermal images) have the attraction of being simpler, cheaper, and more robust. Somewhat unexpectedly trial results show that they also make reflector strips of real value in location; however, the caveat that the strips must be in good condition (see above) still very much applies.

The first electro-optical systems in service will inevitably be 'add-on' installations, but following the military example it is safe to predict that the next 20 years should see fully integrated systems making major impact on overall lifeboat design. Combined with higher transit speeds, the result will be an even more efficient lifeboat service, and even more need to call on their specialised skills in good time to assist in any MOB search.

Chapter 12

EMERGENCY HELICOPTER TRANSFERS

THE LIFEBOAT SERVICE is still vital, but there are many occasions when a helicopter will be used instead, by virtue of its relative speed, and ability to transfer a casualty direct to hospital. For a lifeboat, transfers from a sailing vessel are no different to transfers from a power-driven vessel. By constrast, however, all sailing vessels cause problems for helo crews, and square riggers—by virtue of their size and rig—represent the most severe test of skill, and judgement. Some experience has been gained in exercises with helos, but it must be stressed that this has all been in relatively good conditions, and it is felt that helo operations are generally less well understood than lifeboat operations, hence the very full guidance given here.

In any incident the Master should avoid snap decisions, and take the time fully to discuss all the options with the helo crew. A visit to the local SAR unit is essential personal preparation for this, and both service, and civilian units are always extremely helpful, and interested to discuss various situations as they might apply to a particular vessel. Remember—prior planning prevents poor performance.

Main types of SAR helicopter

Sea King

1. The RN operates various marks of the well known and well proven Sea King, which is a medium-sized helo. In general, the ones seen in a SAR incident will either be ASW aircraft (Mk 6), Utility (Mk 4), or Specialist SAR (Mk 6). The Mk 4 are dark green, have no radar, and will only be seen in daylight in the vicinity of Portland. The Mk 6 is a light grey anti-submarine helo with a good radar and excellent night hover capability, and it also has a specialist SAR version (identified by its bright red nose and tail), which is stripped of all sonar gear and anti-submarine avionics (except the radar), to increase its endurance and carrying capacity.
2. The RN SAR version is the only rescue helo in Europe to employ a SAR diver as part of the crew. Despatched from a 20ft hover the SAR diver has proved his worth on many occasions, rescuing people from otherwise inaccessible positions (underneath capsized boats, for example).
3. The RAF (along with the Norwegians, Germans and Belgians) operate a specialist SAR version which has an inferior radar compared to the RN Sea King, but is otherwise very similar. The RAF SAR helos are bright yellow, and those of the other countries have various other bright colour schemes.
4. The Canadians, Italians and Spanish operate the USN SH-3 type with less capable radar, and inferior night hover capability.
5. The USCG operates its own HH-3 variant, which looks very different as it has a stern ramp. It is a good SAR vehicle operated by highly professional crews.
6. The final version of this ubiquitous helicopter is the civil S 61, which is a specialist SAR helo operated under contract for the British Coastguard, and also by the Royal Danish Air Force. It is larger but otherwise very similar to its military cousin, and it does enjoy a better winching position.
7. Both RN and RAF versions carry a 10-person liferaft, which can be airdropped to survivors.
8. The Sea King is an all-weather helicopter and *in extremis* there is very little that can stop it getting through and achieving its mission.

Merlin (EH 101)

1. This exceptionally capable helo will be replacing the Sea King in the RN, and the Canadian and Italian Navies, from about 1997.
2. Markedly more spacious than the Sea King, it should be an improvement in every way.

Seahawk (SH 60B/F)

1. The maritime version of the Blackhawk, this is replacing the SH-3 in USN service, and there is also a USCG specialist SAR variant.
2. Externally almost as long as the Sea King, this is a fast modern aircraft, but it is very cramped inside and has much less internal volume and carrying capacity.

Lynx

1. This small helicopter is operated by many European navies from their frigates and destroyers.

2. Few versions have a night hover capability, and all have very limited internal space, and a poor winching position. However, it is fast, has a good radar and communications suite (and in some aircraft a thermal imager), and most importantly it may be the only aircraft on the spot.

Dauphin

1. This is the French equivalent of the Lynx, and it has similar capabilities and limitations.

2. The USCG successfully operates a modified specialist SAR version called the **Dolphin**.

Super Jolly Green Giant (MH 53J)

1. This very large helo (it can lift a Sea King) is operated from East Anglia in the combat SAR/special operations role.

2. It is capable of air-to-air refuelling from a KC-130 Hercules and using this has demonstrated the ability to execute a rescue from mid-Atlantic.

3. Overall this is a most impressive aircraft with outstanding equipment (including FLIR), but it is not as maritime-orientated as the other SAR aircraft, and by virtue of its sheer size it is less than ideal for carrying out a pick-up from a confined piece of deck. The rotor downwash is also quite astounding—beware!

Helicopter search and rescue (SAR)

Types of SAR: Assistance that you require can be divided into three main categories:

1. **Medevac/casevac:** This is helicopter assistance to pick up a seriously ill or injured crew member in order to effect a speedy transfer to hospital ashore. This is a most difficult evolution in which the weather, and the patient's condition, are critical factors.

2. **Manoverboard (MOB):** In the circumstances detailed in Chapter 11 you may have requested assistance in a MOB incident. The helo can either participate in a coordinated search, or be autonomous.

3. **Abandon ship:** If as in Chapter 13 you decide to abandon ship, then the helicopter(s) would be purely involved in locating and picking up survivors.

Communications

1. Modern digital avionics have ensured that all maritime helos are capable on both aeronautical and maritime VHF channels, and MF 2182 KHz.

2. When homing to a ship's EPIRB aircraft use the secondary 121.5 MHz VHF signal, because the primary 406.0 MHz UHF satellite uplink signal is just outside their radio coverage. The provision of separate (smaller) 121.5/243.0 MHz locator beacons in all liferaft would be particularly valuable at night, because they would enable the SAR helo(s) to carry out accurate individual homings, and thus save (possibly critical) time, compared to a general search based on the main EPIRB datum.

3. Clear concise communications are the key to a successful operation. It is best that the Captain or the Mate personally man the R/T to avoid the problems of third party discussions.

4. Your initial signal should be in the standard Pan/Mayday format, but if a helo is allocated, then the following additional information should be passed as soon as possible:

   a) Positional accuracy (i.e., Decca, GPS, or DR).

   b) Ship's intended course and speed.

   c) Rig of vessel and masthead height in feet—not everyone will recognise the name, and the implications of the rig may have to be spelt out in very simple terms ('I have horizontal spars projecting 20 ft from each of my two masts').

Hi-Line transfer — a Royal Navy SeaKing helicopter of 771 Naval Air Squadron exercising a hi-line stretcher transfer with an RNLI lifeboat, with the crewman on the starboard quarter correctly tending the hi-line. The trick is to make it look easy, but conditions are not always so favourable, nor are all aircrew equally skilled or experienced (Royal Navy — RNAS Culdrose).

d) For **Medevac:** Any further information on the casualty which may assist in deciding on the hospital, and any special medical requirements. If your diagnosis is based on trained opinion (doctor/nurse) that is useful information, and they may request them to come up in person on the radio circuit.

e) For **MOB:** Information required includes: time and position of MOB datum position; estimated accuracy of that position; location aids deployed (danbuoy, smoke and flame markers, etc.); clothing worn; if wearing lifejacket (and if fitted with a light).

f) For **Abandon Ship:** Total persons on board, total number of liferafts (and how many to be used), aircraft homer aids available (locator beacon frequency) are details that should be passed to the Coastguard.

Time and distance

After you make your radio call the rescue helo will not appear immediately—there is a time factor involved:

- The Maritime Rescue Co-ordination Centre has to activate the SAR unit, and the crew have to start the helo, so you should not expect the longer range (and more complex) aircraft to be airborne in much less than 15 minutes from the time of your call.
- The helo then has to transit to you, and since his average airspeed is about 100 knots, his groundspeed is very wind dependent. For a vessel 35 nm west of the Lizard (Cornwall), with the wind blowing at 30 knots from the west, the nearest SAR helo (based at RNAS Culdrose) would take some 45 minutes to arrive from the time of call-out. This consists of 15 minutes to launch, plus 30 minutes transit time.
- When operating at long range (100 nm) the helo will (rightly) be very concerned about fuel margins, and may have only a short time available on task. Consequently you can greatly improve everything if you give a really accurate position (and update it), and are ready in all respects when he arrives alongside. If you have Decca or Navstar GPS, inform the helo, as the aircraft captain will much appreciate the opportunity to update/check his navigation system.

Medevac preparations

1. The casualty should be dressed warmly, preferably with windproof outer clothing (oilskins)—if going in the liferaft the 'buddy' should be similarly equipped.
2. Stretcher casualties are more complex as the standard shipboard **Neil Robertson** is not acceptable for a lift, and the helo will have to provide its own.
3. Once in direct communication with the helo, pass an update on the patient's condition, your position, and the present weather (including the true windspeed and direction, and the sea state).

Pick-up methods

Any special factors need to be considered. For example, a fixed gaff aft makes the winching area very tight, compared to those vessels which can lower their gaffs. At night it would also help the aircrew if the after masthead, and gaff-peak, were marked with a light, or a cylume stick.

The following methods of pick-up are practical:

- **Hi-line:** The vessel should keep way on, and steer a steady course with the apparent wind from the optimum of Red 20° out to about Red 90° (20° to 90° on the port bow). The wind direction requirement is dictated by the fact that the helo has to hover into wind, has its winching position on the starboard side of the aircraft, and the winch operator acts as the pilot's eyes to talk him into position. To do this he needs a clear view of the vessel (and its obstructions), which means that the vessel must be on the starboard side of the helo. The 'cone of courses' is designed to allow a compromise between the ideal relative wind, and ship motion, and the course chosen will need to be discussed and agreed with the aircrew. Once all this has been sorted out, the helo will come into the overhead at high level (above masthead height), lower a 90-150 ft (27-45 m) natural fibre **hi-line** (messenger), then move clear of the overhead, and lower the winchman. This hi-line is weighted with a lead shot-bag, and is intended as guideline which the deck crew can gather in ready to pull the winchman sideways (into position) for that critical last few feet. The tendency to allow a tug of war to develop must be resisted, and the line must **never** be made fast.

- **Seaboat:** When the helo reports ready the casualty is put in the seaboat, which then moves at least 200m clear, before putting the wind on the port bow (or on the starboard quarter), and steering a steady course for the transfer. The helo can pick-up from the seaboat quite easily (possibly using a hi-line), and well clear of the mother ship. If the seaboat engine is unserviceable, an alternative is to trail it astern on a long line.
- **Liferaft:** When conditions are too rough for the seaboat, a liferaft must be used instead. You should wait until the helo is ready, then put at least one 'buddy' in the raft with the casualty and cut it adrift. The helo will then pick up both casualty and 'buddy,' probably using a hi-line. A liferaft blows around so much in the downwash that accurate spot winching is very difficult, and using the hi-line as a guide greatly speeds the process. This can be prebriefed with the helo.
- **Direct:** The only direct pick-up would be for a survivor actually in the water.

Casualty considerations

The condition of a casualty does have an impact on the evolution, and as far as the helo aircrew are concerned there are two categories:

1. **Walking wounded.** Those cases that are relatively mobile should be picked up with a 'double lift.' This means that the winchman will come down on the wire, attach the survivor into a strop, and then be winched back up attached to the wire along with the survivor. Given the confined areas on board (or in the seaboat) this is obviously the easiest operation.

2. **Stretcher cases.** If the patient needs to be lifted in a stretcher then the whole business becomes more difficult. Your Neil Robertson stretcher can be used for bringing the casualty on deck, but it is not suitable for a helicopter lift, and they will have to lower down their own with the winchman. This is all right when doing a hi-line to the ship, but for the **seaboat** or **liferaft** methods it does create problems:

   (i) **Seaboat:** Here it is probably best to go out and winch down the winchman and the stretcher, return to the ship, put the casualty into the stretcher, and then load everybody back into the seaboat for the final pick-up.

   (ii) **Liferaft:** This is a very difficult situation. You need to put the casualty in your stretcher, load that into a liferaft with at least one 'buddy' and cut it adrift. The helo should have positioned itself in the hover astern of you such that when the liferaft is clear he can move overhead. Once in position he will have to lower down the winchman first, and then the stretcher. The winchman assisted by the 'buddy' has to lift and secure your stretcher (still containing and protecting the casualty) into the helo stretcher, and having completed that little operation the winchman can then be recovered to the helo with the stretcher. There is then a delay while the stretcher is all sorted out in the back of the aircraft, before the winchman can return to pick-up the 'buddy.' The trauma and stress (not to mention the physical pain) for the casualty can be imagined, and you must be sure that it is worth the risk.

Recommended pick-up methods

1. **Daylight—good weather:** Hi-line (only if happy). Seaboat.
2. **Daylight—heavy weather:** Liferaft (if at all).
3. **Night—good weather:** Hi-line (only if crew are experienced). Seaboat.
4. **Night—heavy weather:** Liferaft (if at all).

Hazards

1. **Static:** Static electricity does build up in a helo, but the discharge is most unlikely to be fatal. To be on the safe side aircrew will always attempt to earth the winch hook before it gets near you, and you should avoid touching anything lowered until it has been earthed.

2. **Securing:** You must never secure the winch wire (SWL600 lb), or the hi-line, as that would put the helo at risk, and the aircrew will be forced to cut the winch wire and terminate the rescue.

3. **Winching foul:** There is the risk of the winch wire and or the winchman becoming entangled in your rigging. This has the potential for severe damage to the vessel, and injury to both your crew, and the winchman.

4. **Tip strike:** If aircrew get it wrong and the rotor tips strike the rigging, at best they could just chop through (causing damage and injuries). At worst, a tip strike could cause the helo to crash into you, and the impact of 9 tons of flailing machinery (plus plenty of aviation fuel) would be catastrophic.

5. **Pyrotechnics:** Do not emulate the infamous yachtsman who literally fired at the rescue helo, just as it was making its final approach!

6. **Gash:** Do not ditch gash during helicopter transfers, and ensure that all garbage container lids are secure. The obvious problem is plastic sheeting, or packaging, that can easily get sucked into the engines with unpleasant results—they stop. By contrast, organic waste tends to attract seabirds, and a birdstrike into the engines (or through the cockpit windshield) can be equally effective in terminating helicopter operations.

General points

- **Do not** struggle or interfere while the winchman is putting you in the strop.
- On reaching the aircraft **do not** attempt to climb in yourself; the crew have a tried and tested system, and will do all the work.
- In the aircraft sit exactly where you are told.
- **Do not** inflate your lifejacket in the aircraft.
- Military helos are very spartan and noisy. They have no heating or air-conditioning and vibrate a lot. Survivors are prone to airsickness and emotional distress in such conditions, but a good briefing will help your crew to cope better.

Chapter 13

ABANDON SHIP AND LIFERAFT DRILLS

A LIFERAFT is a singularly unpleasant craft even in fair weather, and in heavy weather it can be positively frightening, even dangerous. The golden rule is that you should stay with the ship until the last possible moment, both because it is far easier to find a ship than a host of small liferafts, and also because until it actually sinks the ship is a safer haven. This was most strongly endorsed by the 1979 Fastnet events when only five of the 24 boats abandoned actually sank; whereas half of the facilities were from those crews who chose the supposed security of a liferaft, only to experience appalling difficulties up to and including complete raft break-up.

Training

Regrettably, few of your professional crew will have been in a liferaft, and even fewer (if any) will have had proper sea survival training. This is a scenario in which the morale and cohesion of your trainee crew will be extremely fragile, and the professional crew will need to project an image of confidence and competence. In fact, they must act much like the crew of a passenger ship in that they are responsible not just for their own survival, but also that of the trainees. Therefore training for the permanent crew is a priority, and dry drills and lectures are no substitute for experience. It is, for example, difficult to appreciate the weight of a 20-man liferaft, and the problems of righting it, if your only experience is a classroom lecture.

Full training should as a minimum cover shore instruction on sea survival and survival equipment, followed by being dropped off a boat at sea, swimming to a capsized liferaft, and carrying out righting and boarding drills before spending at least two hours in the raft, with a helicopter pick-up as the finale. This could all be achieved in a day and a half, and is in reality a small price to pay for the gain in peace of mind.

For on-board instruction for the cadets, good clear pictures are needed of the liferaft both in and out of the container, and of the contents of the raft. Some manufacturers can supply a small model that is inflated with a bicycle pump, and this is a most effective instructional aid.

The points below are all standard, and for a properly trained man (such as a naval aviator) they would be a matter of conditioned reflex. Experience shows that *in extremis* real training (repeated—not once in a lifetime) is what allows you to survive; you must be able to do it blindfold, there is no time to think. The codes of practice at sea provide for the *matériel* side, but the human factor is equally important.

Check list

Before abandoning ship, you should attempt to carry out as many as possible of the following actions:

a) Check that **all** are wearing lifejackets and adequate protective clothing.

b) If there is no time for **Mayday**—activate R/T alarm (2182 KHz), and the Epirb.

c) Have a crash bag ready made up, and also take as many as possible of the following extras to supplement liferaft stocks:
  ● Distress flares.
  ● Distress rockets.
  ● Blankets.
  ● Torches.

d) Designate liferaft commanders, and put your trained crew with weakest cadets.

e) Issue seasick tablets to all (sickness = dehydration).

f) Brief liferaft drill and survival priorities.

g) State that search and rescue units have been alerted. This is important for morale, and morale is one of the keys to survival.

Liferaft drill

A liferaft with a strong trained leader is a place of safety, and its occupants can survive for at least four days with the emergency water supplied. Leaderless and untrained the same people could die within hours. It is therefore vital that all your professional crew are at least familiar with the basics laid out below:

- Board as soon as possible after launching (stability).
  Avoid underloading rafts—use minimum number (location and stability).
  Any underloaded raft—crew stay on weather side (stability).
- Cut the painter and stream the sea anchor.
- Take seasick tablets (if not previously issued).
- Close canopy—but remember to vent for two minutes in 30 (CO_2 build up).
  Bail and sponge out all excess water (insulation).
  Inflate floor (insulation).
  Check for leaks.
  Read instructions inside liferaft.
- Rope rafts together in a string (location).
- Always be prepared for a long wait before rescue.
  Do **not** use water supplies for 24 hours (unless extremely dehydrated).
  Avoid food for three days (except for boiled sweets in raft rations).
- Organise watches (lookout is vital).
- Control use of pyrotechnics—do not waste them. Using distress flares by day, for example, greatly limits their visibility and success rate. Use flares and rockets by night; and smoke and heliograph by day.

Survival priorities

In your lectures to crew and trainees these survival priorites cannot be too highly stressed:

1. **Protection**   Reasonable clothing and the correct operation of the liferaft are the key elements.

2. **Location**   Distress call, Epirb, pyrotechnics, liferaft size and colour, etc., all have their part to play. Strange to relate people have been missed in a liferaft even in the English Channel—do not take your location for granted.

3. **Water** (fresh)   You cannot survive more than three-four days without fresh water, but rations are provided in the liferaft. To make your body function at its optimum you should not drink any of the water ration for the first 24 hours. Obviously this does not apply in the tropics, or if you are injured. The motion of a liferaft is appalling and without pills you will be sick, thus losing valuable fluids and dramatically increasing your water intake requirements. Therefore you **must** take the seasick pills as early as possible, however good your sealegs. All that said, do not miss any opportunity to supplement your ration by collecting rain water, or using a still.

NEVER DRINK SEAWATER or URINE

4. **Food**   You can survive 25-30 days without food, and in the early stages you should avoid solids (the odd Mars bar in your pocket) as this only increases your water intake requirement. Whilst you will suffer some hunger pangs, a week without solid food is relatively easy, and the glucose sweets provided will sustain your energy quite well. (Basic military survival courses stretch to that level with no problems.).

Recovery by helicopter/lifeboat

Time and distance make helicopter recovery the most likely eventuality in Northern European waters, and the techniques are as covered in Chapter 12. Even the Sea King will be unable to recover more than 15-20 at a time, so some sort of a shuttle will be required (probably with several aircraft).

Lifeboats would almost certainly also be involved, and they have good endurance, and survivor capacity. Their only real problem in searches is their relatively low height of eye; consequently the accuracy of your last position report, and the correct use of location aids (flares, etc.), is even more important than for a helicopter.

Provided that your Mayday call included the numbers of people involved, the Coastguard Maritime Rescue Co-ordination Centre (MRCC) should have generated enough aircraft/lifeboats. In fact, everyone would be in overdrive as the loss of a sail training vessel would be regarded as a major emergency.

Chapter 14
DYNAMIC STABILITY AND KNOCKDOWNS

TRADITIONAL big ship stability teaching does not stress external forces at large angles of heel, thus limiting its relevance to sailing vessels, for whom the critical issue is reserve of *dynamic* stability in a knockdown. Stability for large sailing vessels—particularly square riggers—is self-evidently a more complex subject than for power-driven vessels. For example the infamous case of the loss of HMS *Captain* (1870) showed that it is entirely possible for a sailing vessel to have an apparently suitable metacentric height (GM), and to carry sail safely at up to 10-15° heel, but to lack sufficient reserve of stability to cope with the sudden onset of a squall.

Sir William White's classic *Manual of Naval Architecture* remains an excellent source document even though it is long out of print and was last revised in 1894. Since then the main published developments have been largely engendered by the loss of the *Marques* in 1984. This short chapter does not consider yachts, but concentrates on the behaviour and limitations of square riggers and is intended to provoke interest and discussion, rather than to provide definitive answers.

Incidents

The essence of the problem is that while a sailing vessel may operate perfectly safely in 'normal' conditions, squalls have caused the knockdown and loss of a number of vessels. Some confusion has been caused by references to 'microbursts,' but whether titled plain or exotic it is the sudden and dramatic wind increase that is the threat. Commercial ships were particularly vulnerable because of the risk of cargo shift, but HMS *Captain* was a warship, and other well-known knockdown casualties have been pure training ships. The best training ship examples are: HMS *Eurydice* (ship)—off the Isle of Wight in 1878; *Niobe* (jackass barque)—Baltic in 1932; *Marques* (barque)—Atlantic in 1984; and *Pride of Baltimore* (topsail schooner)—Atlantic in 1986. By contrast, TS *Royalist* (brig) suffered a knockdown in Weymouth Bay in 1983, but survived. The Wolfson Unit report lists British sail training vessels, and 'reported' knockdowns, showing that it is a rather more common occurence than one might think, and that no vessel can expect to be exempt.

The fact that *Royalist* is a modern steel design, with yacht-like stability curves extending to over 90°, does not totally explain the difference in outcomes. At least some of the others were large well-found oceangoing vessels that had ridden out many storms. The incidents need to be compared and grouped to detect trends and lessons:

● *Captain.* A controversial and political design, like the airship R101 she was a disaster waiting to happen. She was arrogantly sparred to first class standards, but a glance at Fig 14-1 shows that she had under half the range of static stability enjoyed by her successful turreted contemporary *Monarch*. In fact, this case is a particularly good illustration of how misleading it can be to rely on GM as a guide to stability. Their respective GMs were very comparable, with *Captain's* being slightly larger, and the detailed table extract further highlights the change from apparent comparability of righting lever (GZ) at small angles of heel, where *Captain* was again slightly superior, to the very different picture above 14°, when her lack of freeboard became critical. *Captain* achieved maximum righting lever at half the angle of *Monarch*, and totally ran out of righting moment at an angle which still left *Monarch* with a better GZ than she ever achieved. The fact was that *Captain's* reserve of dynamic stability in heavy weather was quite inadequate, and the squall—though severe—only caused the other vessels in company to lose sails. Downflooding was undoubtedly an additional factor, but she was doomed anyway.

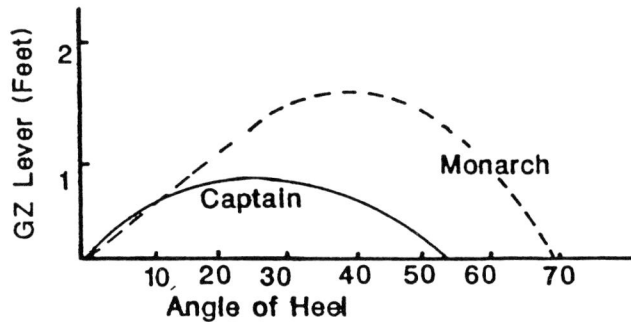

Fig 14-1—Captain *vs* Monarch *GZ*

GZ Lever (inches)	4.0	8.5	10.75*	5.25	zero	minus	*CAPTAIN*	(GM 2.6 ft)
	4.5	8.25	12.25	22.0*	17.5	zero	*MONARCH*	(GM 2.37 ft)
	7°	14°	21°	42°	54.5°	69.5°		

Angle of heel (degrees)

Extracts from Captain *vs* Monarch *GZ curves*

- *Eurydice.* This was a well-proven vessel being driven hard in squally conditions. Despite the conditions her ports were open for ventilation, thus effectively reducing her freeboard from 11 ft to 4 ft and allowing massive downflooding (see Fig 14-2). Without this downflooding—even overdriven as she undoubtedly was—cutting away her rigging progressively should have enabled her to survive or at least kept her afloat long enough for rescue. As it was only two survived (1 AB & 1 OD), and all 200 boys were killed, in a tragedy that inspired Gerard Manley Hopkins to write those chillingly evocative lines of poetry:

> *As half she had righted and hoped to rise*
> *Death teeming in by her portholes*
> *Raced down decks, round messes of mortals.*

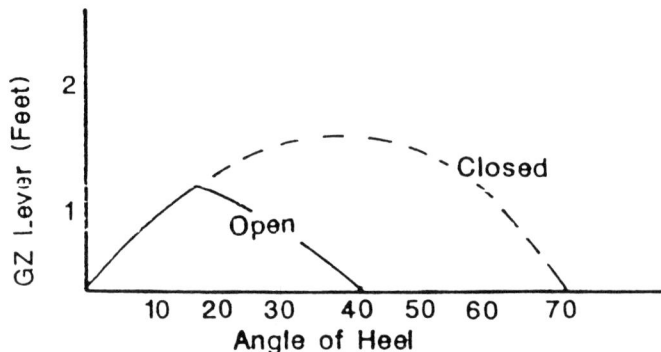

Fig 14-2—Eurydice—*open/closed GZ*

- *Niobe.* A well-found vessel, but she was caught lying virtually becalmed with all her side scuttles open for ventilation. Already in a worst case for knock down the ship side opening had the same effect as in *Eurydice* and sealed her fate.

- *Marques.* This was an elderly vessel with a much modified rig and hull, whose distinctly limited range of stability was unappreciated. She was already in heavy weather when she was laid over by an even stronger gust and suffered downflooding via an open off centre accommodation hatch. In addition she did not possess sufficient operational freeing-ports to clear quickly the water trapped on deck, and therefore possibly had a further free surface effect high up in the vessel, reducing the righting lever still further.

- *Pride of Baltimore.* Newly built—but to a 19th century design—she also had distinctly limited stability and offset hatch. The original 19th century concept was for a deliberately over-sparred vessel, to give maximum light weather performance, putting heavy reliance on the Master for heavy weather survival. In the event she proved unable to respond quickly enough to avoid squall knockdown, with resultant downflooding, capsize and loss.

- *Royalist.* This modern vessel was caught by a sudden squall while hove-to carrying full topsails, in what had been light airs. She was laid right over such that her yard arms were at one stage in the water, but quickly recovered, suffering no structural damage, and insignificant water ingress.

Knockdowns

The theoretical worst case used to consider a vessel heeled to windward, thus giving it additional dynamic energy as it passed through the upright. Conventional wisdom also held that the impact of a squall would in any case cause a greater heeling moment, than a steady wind of the same force, because the sudden application of a force against a resisting object (such as a sailing ship) will cause an initially greater deflection than a steady increase to the same force. As illustrated at Fig 14-3, reserve of dynamic stability is that lying above a given wind force line.

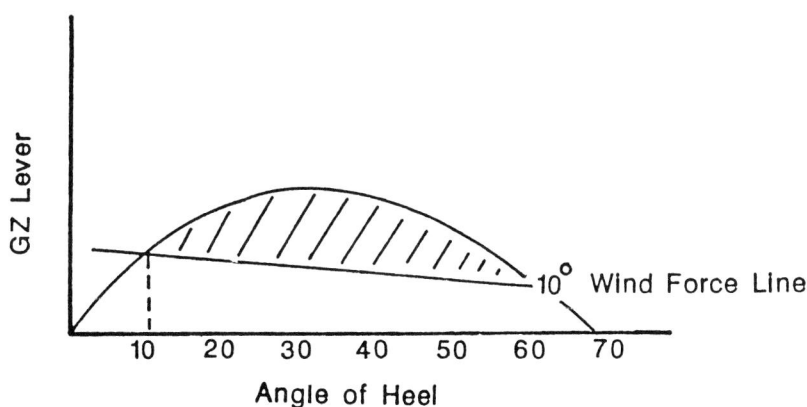

Fig 14-3—Dynamic reserve—generic ship

The mechanisms of what happens were considered very complex but have been simplified by recent Wolfson work. This has found that sailing vessels do have time to start to respond naturally, because these gusts/squalls are not instantaneous, but have a build-up period that (for most vessels) is greater than their period of roll. At their suggestion (as accepted by the DTp for vessels up to 24 m) the stability booklet has diagrams to indicate squall resistance, and to show the sustained heel angle (irrespective of sail set) that allows for squall/gust response before downflooding occurs. This would all appear useful information for the Master in the normal operation of his vessel, and it is to be hoped that this criteria can be extended to larger craft.

The modelling and wind tunnel work done by the Wolfson unit is extremely useful and interesting, but as they recognise there have been only very limited full-scale trials (with limited instrumentation) to validate the many assumptions upon which the modelling is based. A barque and a Bermudan sloop were used, and provided

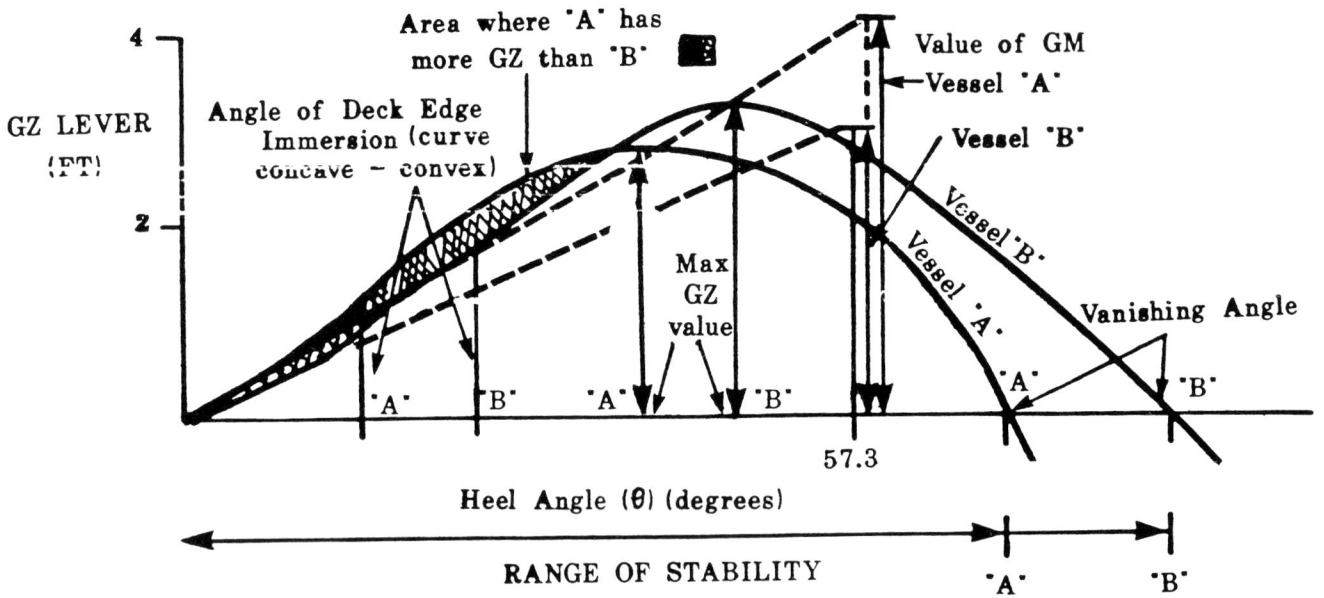

Fig 14-4 GZ Stability curves for contrasting generic vessels.

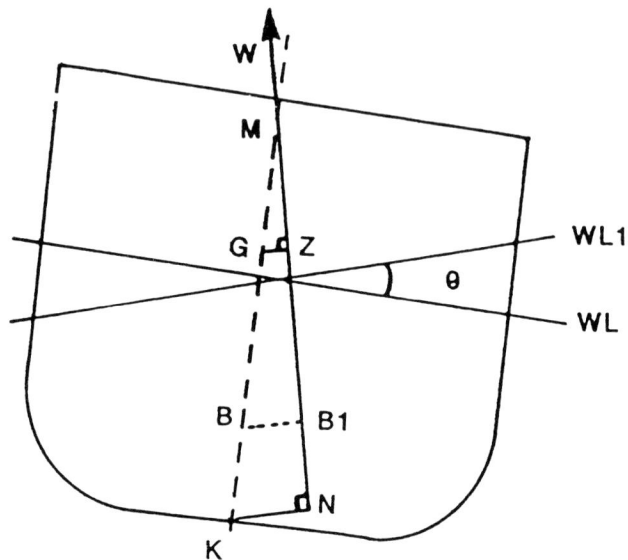

Figure 14-5 Transverse stability diagram.

useful information on the relation between wind strength, sea state, ship speed and angle of heel, but (perhaps fortunately for the crews concerned) the data base was mostly for angles of heel < 30°. When considering the continuing rules of thumb, and modelling reports, the following points still need to be borne in mind:

- The water pressures around the hull, and the air pressures around the rig, both vary over the whole range of operation of a sailing vessel. The relationship between the CE and CLR (as commonly calculated) can only be used for a simple assessment of balance, and they can be misleading labels for the non-specialist.

- The stack of square sails on a mast is a much more intricate aerodynamic structure than is generally appreciated.

- The relationship between wind speed, hull speed, and heeling moment, is very complex, and simplistic assessments can produce apparent inconsistencies with actual experience afloat.

Downflooding

Certain combinations of circumstances will result in knockdown even for the best vessel, and the degree of resultant downflooding is critical to her ability to recover. This is because substantial downflooding causes the additional problems of asymmetric loading and free surface effect, thus making a difficult situation into an irrecoverable one. A good range of dynamic stability is indeed essential, but it is not enough, unless it is combined with maximum resistance to substantial downflooding.

Reports or studies that deal in absolutes and take immersion of a 10 mm vent pipe as the downflooding angle are rather unhelpful, more important is the angle at which 'serious' downflooding occurs via hatches or ports. *Royalist* is a good case in point, having both excellent stability curves and no ship side openings or off-centre hatches, although she does have minor vents and such like that immerse earlier.

The loss of the *Niobe* caused the Royal Navy to revise their rules for sail training ship design (for their aborted 1930s project) to require that the hull was not pierced for opening scuttles (portholes), and the DTp adopted similar rules, which have recently (and rightly) become even more draconian. There can be no half measures, opening scuttles are a proven safety hazard, and vessels so fitted should have them welded irrevocably shut, instead of placing their trust in pre-sea closing down routines.

Excessive water trapped on deck is also a potential problem, as shown in the *Marques* case, and here the recommendation in the code of practice is based on the need to clear the decks in a time close to (or less than) the period of roll. This would seem an eminently sensible design criteria.

Manoeuvres

One factor that is only briefly mentioned in the various reports, and studies, is a vessel's ability to turn rapidly to run off down wind, and a large boomed sail aft certainly impedes bearing away. *Pride of Baltimore* and *Royalist* both attempted to run off, but failed. Nares—amongst others—takes great care to warn that this manoeuvre is very risky, and that unless timed exactly right, attempting to run off often makes matters worse, by bringing the vessel more nearly broadside-to, just at the critical moment of squall impact. It is interesting to note that the technique for hard-driven ocean racers is always to luff up, never bear away; and reading of clipper ship memoirs shows that for them luffing in squally conditions was common practice for the top skippers, whose skill and judgement in using this technique also ensured that their ship did not lose ground to leeward, and thus made a faster passage.

Countermeasures

The normal countermeasures can be split into pre- and post- knockdown, and all have their merits and problems. However the use of engines is seldom mentioned, even though raising steam became a standard precaution in squally weather in the 19th century Navy. Merchant ships being pure sail did not then have that luxury, but that is not an excuse now! The Master of one of the STA schooners reported that when she was knocked down, she was pinned flat with no ability to manoeuvre, until the engineer could start the engines, thus enabling her to be luffed up. There is always the problem of lines over the side, but that seems an acceptable risk given the gain in flexibility (particularly as many vessels now have prop ropecutters). In fact, a more important consideration is the position of the engine intakes, and the ability to run an engine while lying right over on your beam ends.

Two other good points to note are:

- **Square sails:** Whilst it is normal to lower away on the halliards first, in a knockdown that could be fatal, because the harder the wind blows, the firmer the yards will bind, and with the sheets still fast there will be no relief (as was the case in *Captain*). Far better to cast off the sheets, and clew up with the yards still hoisted (see Chapter 4).

- **Crew:** In such circumstances the cry 'All hands on deck' is the instinctive reaction. However, at night it may not be too clever to have half-awake trainees (minus their night vision) spilling out on to the deck, just as you are knocked flat. Consider ordering them to hold below (but ready), until called forward.

Although every sail training ship undoubtedly has its own staff solution, (emergency reference card two at Appendix 1 is an example) no drill can begin to simulate the shock effect that a severe knockdown will have on the whole crew—not just on trainees. Not only does the event occur amazingly quickly, but there may be physical casualties. *Royalist,* for example, had three overboard, of whom two pulled themselves back and one had to be picked up by sea boat, and the Sailing Master (Mate) suffered a broken leg. Any drill that relies totally on key crew members is thus obviously flawed, because in practice these people may be incapacitated, or unavoidably absent (overboard).

Actions after recovery from knockdown

Once the vessel has recovered from the knockdown, you **must** carry out a 100 per cent muster, because in the confusion a manoverboard, or serious injuries, may have been missed. Once that is established, you should recover any menoverboard, attend to the casualties, and assess what immediate repairs are required. This may be quite a list, but your troubles continue even once it is completed.

Such an incident will inevitably get into the Press. Total openness and a proper Press release is most important, as any hint of a cover-up will only result in the worst of ill-informed tabloid hysteria. You must avoid putting yourself in the position of being on the defensive against the Press, because that will ensure a trial by media, and in such cases their verdict is always 'Guilty.'

Conclusion

More and more information is now becoming generally available to the Master in the form of stability booklets and damage booklets (and hopefully more is to come). The Master's 'seat of the pants' judgement remains important, but an informed attitude is essential, both for him and for his OOW. Those officers with professional training in the Navy, or Merchant Service, can build on their already solid grounding in the complex subject of stability. However, yachtsmen coming from the RYA scheme do lack this essential background, and need a short specialised course to raise their knowledge to the required level.

SHIP STABILITY

A COMPREHENSIVE treatment of this subject is outside the scope of this book, but these short notes may usefully complement the foregoing chapter.

Standard stability abbreviations

G Position of the centre of gravity (varies in a non-cargo ship due to stores and some free surface movement of fuel and water).

B Centre of buoyancy of the underwater body (varies with heel angle).

W Displacement (weight and buoyancy are equal and vertically opposite).

M Point where vertical line from 'B' cuts the centre line, and about which the vessel pivots. Valid only for small angles of heel ($< 12°$).

Z Point horizontally opposite 'G'.

K Keel datum point.

N Point horizontally opposite 'K'.

Ø Heel angle.

GM Metacentric height (distance) and only of interest for small angles of heel ($< 12°$).

GZ Righting lever.

KN Distance used for initial calculation of GZ, owing to the movement of 'G' with load state.

CE Geometric centre of above water profile (silhouette) of sails, hull and superstructure, disregarding overlaps and assuming that square sails brace to the centreline.

CLR Geometric centre of underwater hull profile.

Useful formulae

● Righting moment = GZ × W.
● Heeling moment = PAH $\cos^{1,3}\theta$ kg.m.
  P — wind pressure in kg/sq m.
  A — silhouette area (as in CE above).
  H — vertical distance between CE and CLR.

● Wind pressure (P) = $1/50\ V^2$ kg/sq m (working figure).
                    = $1/2\ \varrho V^2$ kg/sq m (architects figure).
  V — wind velocity.
  ϱ — air density.

    Wind pressure is thus related to the square of the velocity, rather than being a simple linear progression, and this factor is regularly underestimated, even when it is understood. It means that the crucial difference between a 40-knot steady wind, and a gust of 56 knots, is not simply the velocity gain of 40 per cent, but that the pressure has increased from 26 to 52 kg/sq m—a rather more dramatic rise of 100 per cent. Similarly an increase in wind strength from 28 knots to 56 knots doubles the velocity, but quadruples the pressure (13 kg to 52 kg). It is more important to appreciate this relationship between pressure and velocity, than it is to be able to parrot mindlessly the velocities for the Beaufort wind scale, and its misleading 'force' terminology.

● Righting Lever GZ = KN — KG sine θ.

Stability curves

The contrasting stability curves at Fig 14-4 are drawn to emphasise the difference between a good range of stability, and basic GM. Vessel 'A' has the better GM and that 'fatally deceptive' initial stiffness shared by both *Captain* and *Marques*, while by contrast vessel 'B,' with less GM, is initially slightly tender, but has a higher freeboard and thus a markedly better range of stability. The curves also illustrate the relationship between various terms such as 'angle of deck edge immersion,' 'angle of maximum righting lever,' 'vanishing angle,' and 'range of stability,' and show how the righting lever varies with the angle of heel.

Regrettably, these important curves tend to remain hidden in a drawer, rather than being prominently displayed, and 'out of sight' is all too often 'out of mind.'

Freeboard

Stability at large angles of heel can probably best be improved by increasing freeboard, provided that this is not too greatly at the expense of GM. It is an apparent anomaly that increased freeboard involves greater structural weight (and increase in draught and displacement); consequently the position of G is raised and GM reduced. This reduction on GM—and initial stability—is the price to be paid for a greater range of stability, increased righting lever, increased angles of deck edge immersion and maximum righting lever, and also a greater reserve of buoyancy.

Considering that a sailing vessel should not be too 'stiff,' regularly experiences relatively large angles of heel, and thus needs a good range of stability, it can be seen that the features that come with higher freeboard are entirely acceptable. Regrettably, Court transcripts of recent accidents show that too many people still view stability in terms of metacentric height, which is depressing when you consider that it is over a century since the *Captain* disaster exposed the fallacy of that position.

Caveats

Lines of statical stability may not precisely reflect a vessel at speed in a seaway, but they are a good guide to the position post-knockdown, when she is virtually stationary. Survival at extreme angles also requires all internal ballast, and all major items of equipment (galley range, etc.), to be totally secure at all angles of heel.

Chapter 15

LEGAL ASPECTS OF WORKING WITH YOUNG PEOPLE

THIS SHORT CHAPTER briefly introduces a subject that is not seamanship in the accepted sense of the word, but is very much part of the Master's business in a sail training vessel with young people. It is all based on English law, but to a great extent the principles should read across to the law in other western countries.

In loco parentis

The Latin tag *in loco parentis* (in place of the parent) is a phrase that is more commonly used than it is understood. The strict legal definition in English law is worth knowing, and is as follows:

> *'The position of a person undertaking the office and duty of a parent to make provision for a child of another.'*

Responsibilities

The Master and the professional crew have this special responsibility to two groups:
- **Children.** Those of compulsory school age (at present (1992) 5-16 years) are classified as children (Education Act 1944).
- **Young persons.** Those aged 16-18 are young persons (Merchant Shipping Act— International Labour Convention—1925).

The key to whether these duties are adequately discharged is 'reasonable care.' This means that the standard of care expected of the Master, acting *in loco parentis*, is that which would be exercised by a 'reasonably careful parent.' He/she can do no more than is 'reasonable' in the particular circumstances.

From the start the organisation/company must decide what standards it wishes from its young people, on board and ashore, and these standards/ guidelines are best spelt out in the pre-joining literature. The actual parents, and their offspring, must be made aware that parental responsibility has 'voluntarily' passed to the Master (as the representative of the company), and that this effectively involves a new parental contract, which may well be more stringent than that at home.

It should also be specified that if the trainee fails to comply with the standards set— despite the best efforts of the Master—then the Master/company will not be liable, and that excessively disruptive behaviour, or failure to obey the Master's lawful instructions (or those of his officers), could result in the trainee being landed early. This latter action would never be taken arbitrarily, but must be an option available to the Master, who has a responsibility not just to the individual, but to the other trainees, and the crew as a whole.

It may sound obvious, but the parents should be made aware of the fact that their offspring will be subject to foreign laws when in foreign ports.

Problem areas

The following potential problem areas have been identified for which the Master has a special responsibility towards those under 18, though in some cases the problems are far from exclusively limited to them (drink, drugs, etc.).
- **Age of consent.** This is not a problem on board, owing to the complete lack of privacy for the trainees, but it could be a problem abroad because countries have different break-point ages, some as low as 12.
- **Crew/trainee relations.** Obviously this is illegal for under 16, but 16-18-year-old females are emotionally vulnerable, however legal. Here the Master will need to exercise parental judgement, as what might be acceptable for a fellow trainee, or a 19-year-old boatswain's mate, would be totally unacceptable in a chief Mate aged 34. The simplest rule is that in Chapter 10, where **all** trainees are 'off limits' to the crew. This may sound rather draconian to male readers, but many females are (understandably) even more outspoken on the subject of middle-aged seaborne lotharios and sail training.

- **Smoking.** This may be a lost cause to some extent, but certainly cigarettes, etc., cannot be supplied from the ship's store to under-age trainees, as you are as legally bound as a storekeeper ashore. Most vessels so severely restrict smoking—none below decks (safety), only certain upper deck areas (cleanliness), etc.—that they have a much firmer grip than the average school, or parent.
- **Alcohol and alcohol abuse.** Under-age drinking on board is easily avoided, but ashore all you can do is remind them of their status, and try to persuade the over 18s not to encourage the all too prevalent belief that drinking to excess is a sign of maturity. Heavy drinking in teenagers, and young people generally, is a growth area, and it is important to emphasise that if someone does drink too much, then they must not be abandoned to their fate by their fellow trainees (however obnoxious they may be at the time), because that is when real tragedies occur.
- **Drug and substance abuse.** No matter how young your trainees, nor how good their backgrounds, these related problems must always be considered. Apart from controlled drugs, other substances (solvents, glue, etc.) also fall under the description of drug abuse, and the whole frightening business is a constantly evolving scene. In one vessel they started by thinking that they had a leaking gas system, only to find that one of their trainees was taking massive doses to get high!! At sea these people put not only their own lives at risk, but can also seriously endanger everyone. In order to understand properly the problem, spot the addict, and protect the remainder, the Master (at least) needs to have attended some sort of course, or teach-in, on the subject. Naval personnel are fortunate in this respect, since they will all have had the mandatory lectures/films during their service, and the officers (and many senior rates) will have attended the excellent courses run by the RN drugs squad.
- **Bullying.** There is some justice in this world, and generally the obvious bully boys seem to suffer from seasickness to a greater extent than their potential victims, and their 'street cred' is further eroded by their obvious lack of enthusiasm for work aloft at sea. However, there are exceptions, and if they are unduly disruptive, and do not respond to firm treatment, and close supervision, then for the greater good, the ultimate sanction of landing, and return home, may have to be invoked.
- **Arrest.** In the event of a young trainee being arrested for a crime allegedly committed ashore (in UK or abroad), then the Master would have responsibilities over and above the normal, and would need to act *in loco parentis* either until the case was resolved (simple case), or until the parents could be informed and take over responsibility.
- **Illness/injury.** If a trainee has to be admitted to hospital for any cause, then once again the Master remains *in loco parentis*, until such time as the parents can take over.
- **Medical conditions.** Well-intentioned parents (amazingly sometimes 'advised' by a doctor) have been known to deliberately fail to declare conditions such as vertigo in the parents' consent form, on the grounds that 'it might affect the way that they are treated'!! A positive verbal double check on joining is therefore a valuable precaution that has paid dividends in the past.

Shore leave

In many ports it may be necessary to advise trainees to stay in groups of at least three, certainly after dark. The Master has responsibility in this respect to all trainees, but those under 18 warrant special briefing, and a sensible approach which avoids 'nannying.' Certainly all night leave should not be extended to them, and some form of curfew imposed, however unpopular it may be. The youngsters will undoubtedly initially protest (possibly truthfully) that their parents allow them more license, and that the Master is being cruel, and heartless. However, later on these same people will often admit that the curfew got them out of potentially awkward situations.

Crew exchanges

These exchanges are a popular feature of the Tall Ships Races, and no-one would wish to prevent the younger trainees from taking part. However, joining instructions need to make it clear that this activity is on the agenda. Furthermore the Master will have to take reasonable care to establish the suitability of any vessel run by an individual (or organisation) that is not well established. The potential new Master, and the vessel, should meet the standards that the trainee's parent(s) signified by entrusting the child to the care of the original Master.

Appendix 1
EXAMPLES OF EMERGENCY REFERENCE CARDS

THE FOUR CARDS which follow were produced for a brig, and illustrate the level of information that can be put into a simple ready reference format. If used by other vessels, suitable amendments would have to be made to account for their special characteristics:

CARD ONE

MANOVERBOARD — UNDER SAIL

REMEMBER — 1 min delay at 6 kn = 200yd from MOB.

a. Fair Weather: (< 20 kn wind speed).

Emergency Stop.

Throw the vessel aback to stop her — then sort yourself out.

In Visual contact:
Remain Hove-to and recover by seaboat, or heaving line.

Lost contact: (Probably be at night).
Furl all (square) sail and check no lines in the water.
Lower and man seaboat — keep it close (ready to pounce).
Conduct a square search starting on the datum.
Station your better cadets in the best lookout positions.

b. Heavy Weather:
Emergency stop is out of the question.
Undoubtedly lose contact with the MOB.
Launching seaboat only puts others at risk — ship pick-up only option.
Motor-sailing to a 'modified' square search may be best.
Accurate search will be very difficult.
(MOB must reach lifcraft to have a good chance of recovery or survival)

MANOVERBOARD — UNDER POWER

All Weather:
Textbook WILLIAMSON turn.
(adjust roll out to compensate for wind)

CARD TWO

KNOCK DOWN COUNTERMEASURES

a.   Pre Knock down:

   (i)   Run off (if time — risk is being caught broadside on).

   (ii)   Luff up (requires exceptional judgement — risks dismasting).

   (iii)   Take in all possible sail (leave yards hoisted).

   (iv)   Close down *ALL* hatches and scuttles (including superstructure).

   (v)   Close *ALL* wateright doors (evacuate Forepeak mess).

   (vi)   Move hands already on deck to weather side — clip on lifelines.

   (vii)   Start engines (stop weather engine on impact).

   (viii) Hands below decks muster on mess deck and kit up.

b.   Post Knock down:

   (i)   Take in all remaining sail (leave yards hoisted).

   (ii)   Start leeward engine.

   (iii)   Use power to run off — or luff up (as appropriate).

   (iv)   Start mechanical bilge pumps.

   (v)   Treat injured, and recover any MOB.

   (vi)   Carry out 100% muster (*ESSENTIAL*).

   (vii)   Assess damage.

CARD THREE

HELICOPTER CASEVAC/MEDEVAC

General: Take the time to fully discuss all the options with the helo.
The fixed gaff makes the winching area aft very tight.

Pick up Methods:

a. Hi-Line: A long weighted messenger (Hi-Line) is lowered from the overhead then the Helo backs off and lowers the winchman. The Hi-line is used to pull the winchman in the last few feet.

b. Seaboat: When the helo is ready put the casualty in the boat. It should then move 200 yd clear, put the wind on the port bow and steer a steady course. The helo will pick up direct (possibly using the Hi-Line).

c. Liferaft: Put at least one 'buddy' in the raft with the casualty, then cut adrift. The helo should then pick up both casualty and 'buddy.'

Recommended Methods:

a. Daylight — Good Weather: Hi-Line or Seaboat.

b. Daylight — Heavy Weather: Liferaft.

c. Night — Good Weather: Seaboat (Hi-Line only in perfect conditions).

d. Night — Heavy Weather: Liferaft.

HAZARDS:

a. Static   —   Avoid touching anything lowered until earthed.

b. Securing   —   NEVER secure the winch wire (SWL 600lb) or the Hi-Line.

c. Winching Foul   —   Risk of the winch wire and/or the winchman becoming entangled in your rigging.

d. Tip Strike   —   If the rotor tips strike the rigging they could just chop through (causing damage & injuries), or cause the helo to crash into you.

e. Pyrotechnics   —   Do NOT fire anywhere near the helo.

CARD FOUR

ABANDON SHIP

General: Stay with the ship — if at all possible.

Check List:

a.   ALL to wear lifejackets & adequate protective clothing.

b.   Activate R/T alarm (2182 KHz), and the EPIRB.

c.   Take the following extras:

    (i)   Distress flares   (12 — boxed)   (Emergency locker)

    (ii)   Distress rockets   (12 — boxed)   (Emergency locker)

    (iii)   Crash bag   (one)   (Emergency locker)

    (iv)   Torches   (as many as possible)

d.   Designate at least one adult per Liferaft (trained crew with weakest).

e.   Issue seasick tablets to ALL (sickness = dehydration).

f.   Brief liferaft drill & survival priorities.

g.   State that SAR units en-route (morale).

Liferaft Drill:

a.   Board as soon as possible after launching (stability).

b.   Avoid underloading rafts — use minimum number of rafts (location).

c.   Once cut loose — stream sea anchor.

d.   Take seasick tablets (if not previously issued).

e.   Any underloaded raft — crew stay on weather side (stability).

f.   Close canopy — but remember to vent for 2 min in 30 (CO_2 build up).

g.   Rope rafts together in a string (location).

h.   Do NOT use water supplies for 24 hrs — unless extremely dehydrated.

i.   Avoid food for 3 days — except for boiled sweets in raft rations.

j.   Always be prepared for a long wait before rescue.

k.   Organise watches (lookout vital).

l.   Read survival handbook (one in every raft).

Survival priorities:

1.   Protection   — Clothing & liferaft.

2.   Location   — Distress call, EPIRP, pyrotechnics, liferaft.

3.   Water (fresh)   — Cannot survive more than 3-4 days without.

                      NEVER DRINK SEAWATER or URINE

4.   Food   — Can survive 25 — 30 days without.

Appendix 2

REFERENCES AND RECOMMENDED READING

THE DIFFICULTY in any book on seamanship is that there is a certain element of interpretation, and also the author's character and experience has an impact on the balance of the book. Thus there is a need to consider other views and also to look at the primary sources. This is not always easy with square rig seamanship as so many of the sources are now so long out of print that they are expensive collectors items (marked * in the list). However, these old books are worth seeking out as they were written by highly experienced practitioners of the art, such as the redoubtable Victorian arctic explorer, and hydrographer, Vice-Admiral Sir George Nares.

The resultant list is not exhaustive, but it does provide a reasonable guide to the variety of publications that are relevant, and may be useful for those preparing for examination.

Square rig and seamanship textbooks

1. R.M. Willoughby *Square Rig Seamanship* (1989).
2. P.M. Regan/E.H. Daniels *Eagle Seamanship* (1979 or 1990 edition).
3. HMSO *The Admiralty Manual of Seamanship* (Vols 1-4) (latest edition).
4. G. Danton *Theory & Practice of Seamanship* (10th edition 1988).
5. *G.S. Nares *Seamanship* (7th edition 1897) (2nd edition 1862—reprinted 1979)*.
6. *A.H. Alston *Seamanship* (4th edition 1902)*.
7. *W. Hutchinson *Practical Seamanship* (1st edition 1777—reprinted 1979)*.
8. J. Harland *Seamanship in the Age of Sail* (1985) (a useful compendium albeit by a historian rather than a seaman).
9. C.W. Ashley *The Ashley Book of Knots* (1944—regularly reprinted).

Heavy weather seamanship

1. K. Adlard Coles *Heavy Weather Sailing* (4th edition 1991).
2. W. Kotch & R. Henderson *Heavy Weather Guide* (2nd edition 1984).
3. C. van Rietschoten & B. Pickthall *Blue Water Racing* (1985).

Ship design and rigging

1. *W.H. White *Manual of Naval Architecture* (3rd edition 1894)*.
2. *J. Fincham *A Treatise on Masting Ships & Mast Making* (3rd edition 1854—reprinted 1982)*.
3. H.A. Underhill *Masting and Rigging* (1946—reprinted regularly).
4. J. Lees *Masting and Rigging of English Ships of War* (2nd edition 1984).
5. Germanischer Lloyd *Rules for the Masting & Rigging of Sailing Ships* (latest edition).
6. M. Sheahan *Sailing Rigs & Spars* (1990).
7. J. Howard-Williams *Sails* (1988).

Passage planning

1. HMSO *Ocean Passages of the World* (4th edition 1987) (the sailing routes section remains unrivalled).
2. HMSO Chart 5309 *Tracks followed by sailing and low-powered steam vessels* (shows many additional routes to the above volume).

Shiphandling and miscellaneous

1. R.S. Crenshaw *Naval Shiphandling* (4th edition 1988).
2. J.A.H. Paffett *Ships and Water* (1990).
3. H.M.S.O. *The Mariners Handbook* (6th edition 1989).
4. A. Watts *Cruising Weather* (1982).
5. N. Calder *Repairs at Sea* (1988).

Sail training and the Tall Ships Races

1. J. Hamilton *Sail Training—The message of the tall ships* (1988) (the gospel according to the STA Race director).
2. J. Atkinson *A Girl in Square Rig* (1988) (a trainee's view of the 1976 trans-Atlantic Tall Ships Races).

Classics

1. J.S. Learmont *Master in Sail* (1954) (contains possibly the best account of cutting away masts and rigging).
2. W.A. Robinson *To the Great Southern Sea* (1953) (storm survival in a small brigantine).
3. A.J. Villiers *The Way of a Ship* (1953—reprinted regularly).
4. A.J. Villiers *The Cruise of the Conrad* (1938—reprinted regularly) (round the world with a young trainee crew).
5. G.P. McGowan *The Skipper and the Eagle* (1948) (*Eagle's* post-war delivery trip via the tail of a hurricane).
6. A. Seligman *The Voyage of the Cap Pilar* (1939) (round the world in an ex-Grand Banks barquentine).

Regulations

1. A.N. Cockroft and J.N.F. Lameijer *A Guide to the Collision Avoidance Rules* (4th edition 1990).
2. DTp *The Safety of Sail Training Ships—A Code of Practice <24m* (1990).
3. DTp *Model Stability Booklet for Sail Training Ships <24m* (1990).

Reports of interest

1. DTp *The Auxiliary Barque Marques—DTp Report of Court Number 8073* (1987).
2. US National Transportation Safety Board *Marine Accident Report—Capsizing & Sinking of the US Sailing vessel Pride of Baltimore* (1987).
3. University of Southampton Wolfson Unit *DTp Report Number 798* (1987).
4. B. Deakin (Wolfson Unit) *The development of stability standards for UK sailing vessels* RINA (Spring 1990).

Historical studies

1. J. Leather *Gaff Rig* (2nd edition 1989).
2. B. Greenhill *The Merchant Schooners* (1988 edition).
3. D.C. MacGregor *Merchant Sailing Ships 1850-75* (1984).
4. A.H. Moore *Last days of Mast & Sail* (1925—reprinted 1970).
5. D.K. Brown *Before the Ironclad* (1990).
6. G.A. Ballard *The Black Battlefleet* (1980).
7. F.G. Carr *Medley of Mast & Sail—1* (1976).
8. A.A. Hunt *Medley of Mast & Sail—2* (1981).

Appendix 3
USEFUL GERMAN AND NORWEGIAN SAILING TERMS

ENGLISH	GERMAN	NORWEGIAN
Avast	Halt	Hold an
Belay	Belegen/Festmachen	Sett fast
Belaying Pin	Karvell/Nagel	Koffernagle
Boom	Baum	Bom
Bowsprit	Bugspriet	Baugspryd
Braces	Brassen	Braser
Brails	Gaffel gordinger	Briller
Buntline	Gording	Gording
Clewline	Geitau	Gitau
Come-up	Komm auf	Kom opp
Davit falls	Davit Fall	Davider
Downhaul	Niederholder	Nedhauler
Ease	Fieren	Fir/Slakk
Footrope	Fußpferd	Fotperd
Furl	Beschlagen	Besla
Gaff	Gaffel	Gaffel
Gasket	(Beschlag) zeising	Seising
Gantline	Toppjolle	Toppjoller
Halliard	Fall	Fall
Haul	Holen	Hal
Jackstay	Jackstag	Jackstag
Leechline	Nock-Gordinge	Nokk-Gording
Lift(s)	Toppnant (en)	Topplent
Loose (to)	Losmachen	Brek ut
Mast	Mast	Mast
Manoverboard	Mann über Bord	Mann over bord
Preventer	Bullentaljen	Preventer
Sail	Segel	Seil
Sheet	Schot	Skjot
Stopper	Stopper	Stopper
Stay	Stag	Stag
Tack	Hals	Hals
Vang	Gaffelgeer/gei	Gaffelgjerder/gai
Veer	Auffieren	Slakk
Windlass	Ankerwinsch/spill	Gangspil
Yard	Rah	Ra
To Tack	Wenden	Vende
To Wear	Halsen	Halse
To Boxhaul	Back-halsen	Back Halse
To Heave-to	Beidrehen	Dreietil/Legge bi
Foresail	Fock (segel)	Fokk (seil)
Mainsail	Großsegel	Storseil
Crossjack	Bagien	Bergine
Topsail (Upper)	Mars (ober)	Merse (over)
(Lower)	(unter)	(under) or Stump
Topgallant	Bram	Bram
Royal	Royal	Royl
Flying Jib	Jager	Jager
Jib (inner)	(Innen) Klüver	(Indre) Klyver
(outer)	(Aussen)	(Ytre)
Staysail	Stagsegel	Stagseil
Spanker	Besan	Mesan
Fore	Vor	For
Main	Gross	Stor
Mizzen	Besan/Kreuz	Mesan/Krys
Port	Backbord	Barbord
Starboard	Steuerbord	Styrbord

Appendix 4

LIST OF ABBREVIATIONS

AB	Able Seaman
ARPA	Automatic Radar Plotting Aids
Avcat	Aviation fuel (basically kerosene)
BA	Breathing Apparatus (compressed air)
Casevac	Casualty evacuation (by helo) of injured to shore hospital
CE	Centre of Effort
CLR	Centre of Lateral Resistance
CO_2	Carbon dioxide
CPP	Controllable Pitch Propeller
Datum	Reference position (as in chart datum)
DTp	Department of Transport
EPIRB	Emergency Position Indicating Radio Beacon
ETA	Estimated Time of Arrival
FLIR	Forward Looking Infra-Red
FSWR	Flexible Steel Wire Rope (running rigging) (see SWR)
ft	feet
Gash	Any waste material (both organic and non-organic)
GHz	Giga-Hertz (unit of frequency)
GM	Metacentric height
GZ	Righting lever
Helo	Helicopter
IMO	International Maritime Organization
(jg)	junior grade (as in lieutenant jg USN)
kg	kilogram
KHz	Kilo-Hertz (unit of frequency)
kn	knot(s) (speed)
m	metres
Marpol	IMO Regulations for control of garbage pollution at sea
Medevac	Medical evacuation (by helo) of sick/ill to shore hospital
min	minutes
MF	Medium Frequency (radio)
MHz	Mega-Hertz (unit of frequency)
MOB	Manoverboard
MRCC	Maritime Rescue Co-ordination Centre
Mk	Mark (as in Sea King Mk 6)
nm	nautical miles
NUC	Not Under Command
OD	Ordinary Seaman
OOW	Officer of the Watch
PLB	Personal Locator Beacon (aircrew type mini-EPIRB)
RN	Royal Navy
RNAS	Royal Naval Air Station
RNLI	Royal National Lifeboat Institution
RYA	Royal Yachting Association
RAF	Royal Air Force
Radhaz	Radio/Radar electro-magnetic radiation hazard
R/T	Radio-Telephone

SAR	Search and Rescue
SWL	Safe Working Load
SWR	Steel Wire Rope (standing rigging) (see FSWR)
SSWR	Stainless Steel Wire Rope
SS	Stainless Steel (fittings etc.)
SOLAS	Safety of Life at Sea (International Convention)
sq ft	square foot/feet
sq m	square metre/metres
TV	Television
UK	United Kingdom
USN	United States Navy
USNR	United States Naval Reserve
USCG	United States Coast Guard
U/S	Unserviceable
UV	Ultraviolet (radiation)
UHF	Ultra High Frequency (radio)
VHF	Very High Frequency (radio)
VLCC	Very Large Crude Carrier
VPP	Variable Pitch Propeller
Vs	Versus
WL	Waterline
yd	yard(s)
7×7	Type of SWR — 7 strands with 7 wires each
7×19	Type of SWR — 7 strands with 19 wires each
1×19	Type of SWR — 19 thick wires with no strands

Note

Aircraft variant identifying letters and numbers (SH-3 etc.) are included in the text to differentiate between such variants, but the arcane logic behind the system is omitted due to lack of space!

THE NAUTICAL INSTITUTE

The Nautical Institute is an international professional body for qualified mariners whose principal aim is to promote a high standard of knowledge, competence and qualifications amongst those in control of seagoing craft.

The Institute publishes a monthly journal *SEAWAYS* and is actively involved in promoting good operational practices as demonstrated by this book.

Other projects and certificate schemes include The Nautical Institute on Command, The Work of The Nautical Surveyor, The Work of The Harbour Master, Pilotage, and Management.

There are now some 6,000 members in over 70 different countries; the requirements for full membership are a master's foreign-going certificate from a recognised administration or naval ship command qualifications.

For more information and an application form, write to The Secretary, The Nautical Institute, 202 Lambeth Road, LONDON SE1 7LQ, UK, or telephone 071-928 1351.